Student Solutions Manual
for

Applied Regression Analysis
A Second Course in Business and Economic Statistics

Fourth Edition

Terry E. Dielman
Texas Christian University

BROOKS/COLE
CENGAGE Learning

Australia • Brazil • Japan • Korea • Mexico • Singapore • Spain • United Kingdom • United States

Student Solutions Manual for Applied Regression Analysis, Fourth Edition
Terry E. Dielman

© 2005 Brooks/Cole, Cengage Learning

ALL RIGHTS RESERVED. No part of this work covered by the copyright herein may be reproduced, transmitted, stored or used in any form or by any means graphic, electronic, or mechanical, including but not limited to photocopying, recording, scanning, digitizing, taping, Web distribution, information networks, or information storage and retrieval systems, except as permitted under Section 107 or 108 of the 1976 United States Copyright Act, without the prior written permission of the publisher.

> For product information and technology assistance, contact us at
> **Cengage Learning Academic Resource Center, 1-800-423-0563**
> For permission to use material from this text or product,
> submit all requests online at **www.cengage.com/permissions**
> Further permissions questions can be emailed to
> **permissionrequest@cengage.com**

ISBN-13: 978-0-534-46550-6

ISBN-10: 0-534-46550-1

Brooks/Cole
10 Davis Drive
Belmont, CA 94002-3098
USA

Cengage Learning products are represented in Canada by Nelson Education, Ltd.

For your course and learning solutions, visit **academic.cengage.com**

Purchase any of our products at your local college store or at our preferred online store **www.ichapters.com**

Printed in the United States of America
3 4 5 6 7   13 12 11 10 09
ED167

Table of Contents

Preface...		iv
Chapter Two:	Review of Basic Statistical Concepts................................	1
Chapter Three:	Simple Regression Analysis..	11
Chapter Four:	Multiple Regression Analysis...	37
Chapter Five:	Fitting Curves to Data..	55
Chapter Six:	Assessing the Assumptions of the Regression Model...........	63
Chapter Seven:	Using Indicator and Interaction Variables...........................	99
Chapter Eight:	Variable Selection...	137
Chapter Nine:	An Introduction to Analysis of Variance.............................	149
Chapter Ten:	Qualitative Dependent Variables: An Introduction to Discriminant Analysis and Logistic Regression.................	157
Chapter Eleven:	Forecasting Methods for Time Series Data........................	165
Appendix A:	Summation Notation...	169
Appendix D:	Matrices and Their Application to Regression Analysis...........	169

Preface

Throughout this solution manual, output from MINITAB has been used in all problems where computer analysis is required. The output should be similar to that which would have been produced had Excel, SAS or some other software been used.

Also, some problems have less than complete answers. These are primarily problems where regression model-building is required. Often in these cases, suggestions have been made on which choice of variables might be appropriate or which assumptions might be violated, but a final "right answer" may not be given. There may be several competing models which could be equally acceptable. Even where I have provided a suggested model, there are instances where an alternative model that is nearly as good as the one suggested might be developed. (I always stress nearly as good. Because I am the author, mine must be better, but other solutions can be nearly as good.) There is not necessarily one right answer to the model-building problems.

Note that the solutions begin with Chapter 2. There are no problems in Chapter 1.

Two other comments are in order concerning the solutions:

When test statistics follow the t distribution, I use t values from the table in the text as long as the degrees of freedom(df) for the problem are 30 or less (df $\leq$30). When df > 30, I use z values. There are other options when df > 30: 1) The exact t value could be used if it was available. For example, there are extended t tables that could be used to provide the exact value or MINITAB could be used to determine the exact value; 2) The table in the book could be used to provide an approximation to the exact t value. Some slight difference in results is to be expected if approximate or exact t values are used in place of z values, but the substance of the solutions should remain the same. Of course, for problems done with a statistical package such as MINITAB, the exact t values are used automatically whenever a t value is appropriate.

For some problems, rounding may cause answers to differ from those in the solutions manual. For example, in problem 1 of Chapter 3, the value of the intercept coefficient is computed as 7.1001. This answer may differ slightly due to rounding. Rounding may also affect answers when using MINITAB output. In problem 25 of Chapter 3, the regression equation could be expressed as CONS = 2521 + 0.827INCOME or as CONS = 2521 + 0.82690INCOME depending on the extent of rounding desired.

CHAPTER TWO
Review of Basic Statistical Concepts

2.1 Descriptive Statistics: HWYMPG

```
Variable    Mean    StDev   Median
HWYMPG      28.150  6.534   28.000
```

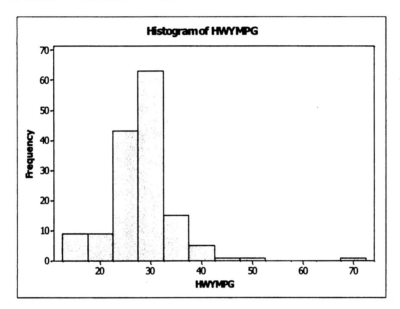

2.3

x	f(x)	xf(x)	x^2	$x^2 f(x)$
1	1/6	1/6	1	1/6
2	1/6	2/6	4	4/6
3	1/6	3/6	9	9/6
4	1/6	4/6	16	16/6
5	1/6	5/6	25	25/6
6	1/6	6/6	36	36/6
		21/6		91/6

$\mu = E(X) = 21/6 = 3.5$

$\sigma^2 = \text{Var}(X) = \sum x^2 f(x) - \mu^2 = 91/6 - (21/6)^2 = 2.917$

$\sigma = \sqrt{2.917} = 1.71$

2.5

x	f(x)	xf(x)	x^2	$x^2 f(x)$
2	1/36	2/36	4	4/36
3	2/36	6/36	9	18/36
4	3/36	12/36	16	48/36
5	4/36	20/36	25	100/36
6	5/36	30/36	36	180/36
7	6/36	42/36	49	294/36
8	5/36	40/36	64	320/36
9	4/36	36/36	81	324/36
10	3/36	30/36	100	300/36
11	2/36	22/36	121	242/36
12	1/36	12/36	144	144/36
		252/36		1974/36

$\mu = E(X) = 252/36 = 7$

$\sigma^2 = \text{Var}(X) = \sum x^2 f(x) - \mu^2 = 1974/36 - (252/36)^2 = 5.83$

$\sigma = \sqrt{5.83} = 2.415$

2.7

x	f(x)	xf(x)
0	0.55	0.00
1	0.15	0.15
2	0.10	0.20
3	0.10	0.30
4	0.05	0.20
5	0.05	0.25
		1.10

a $\mu = E(X) = 1.10$

b ($100)(1.10) = $110.

c $P(X > 2) = 0.1 + 0.05 + 0.05 = 0.2$

2.9

a $P(X \geq 700) = P\left(Z \geq \dfrac{700-800}{100}\right) = P(Z \geq -1) = 0.3413 + 0.5 = 0.8413$

b $P(X \geq 1000) = P\left(Z \geq \dfrac{1000-800}{100}\right) = P(Z \geq 2) = 0.5 - 0.4772$

   $= 0.0228$

c $(2500)(0.8413) = 2103.25$ (approximately 2103)

2.11 **a** $P(-z < Z < z) = 0.9544$
      $P(0 < Z < z) = 0.4772$ so, $z = 2.0$

  b $P(-z < Z < z) = 0.9010$
      $P(0 < Z < z) = 0.4505$ so, $z = 1.65$

  c $P(-z < Z < z) = 0.9802$
      $P(0 < Z < z) = 0.4901$ so, $z = 2.33$

  d $P(-z < Z < z) = 0.9902$
      $P(0 < Z < z) = 0.4951$ so, $z = 2.58$

2.13 $P(z < Z < 0) = 0.48$ so, $z = -2.06$ (approximately)

  $k = 84 - 2.06(7) = 69.58$ The number of months for the warranty should be 69.58 (around 69 or 70 months).

2.15 **a** $P(\overline{X} > 2220) = P\left(Z > \dfrac{2220 - 2200}{50/\sqrt{25}}\right) = P(Z > 2.0) = 0.5 - 0.4772$

  $= 0.0228$

  b $P(-10 \le \overline{X} - \mu \le 10) = P\left(\dfrac{-10}{50/\sqrt{25}} \le Z \le \dfrac{10}{50/\sqrt{25}}\right)$

  $= P(-1 \le Z \le 1) = 0.3413 + 0.3413 = 0.6826$

  The probability that the sample mean is within $10 of the population mean is 0.6826, so the probability that it differs by more than $10 is $1 - 0.6826 = 0.3174$.

2.17 **a** $P(X > 200.2) = P\left(Z > \dfrac{200.2 - 200}{0.1}\right) = P(Z > 2) = 0.5 - 0.4772$

  $= 0.0228$

  b $P(\overline{X} > 200.2) = P\left(Z > \dfrac{200.2 - 200}{0.1/\sqrt{25}}\right) = P(Z > 10) = 0.0$

  c Yes, because the population is normally distributed. This ensures that the sampling distribution of the sample mean will also be normal, even if the sample size is small.

2.19 $P(-1 \leq \bar{X} - \mu \leq 1) = P\left(\dfrac{-1}{5/\sqrt{16}} \leq Z \leq \dfrac{1}{5/\sqrt{16}}\right) = P(-0.8 \leq Z \leq 0.8)$

$= 0.2881 + 0.2881 = 0.5762$

2.21 One way to approach this problem would be as follows. Suppose the average lifetime of the new hard drives has not changed so it is still 3250 hours. Are the data obtained consistent with this stated average lifetime? If so, then a sample mean of 3575 hours in a random sample of 50 hard drives should not be unusual. To measure how unusual this value is, find the probability of obtaining a sample mean of 3575 or more if the true average lifetime of the hard drives is 3250.

$P(\bar{X} > 3575) = P\left(Z > \dfrac{3575 - 3250}{700/\sqrt{50}}\right) = P(Z > 3.28) = 0.5 - 0.5 = 0.0$

If the mean lifetime of the hard drives has not changed, then the probability of finding a sample mean lifetime of 3575 hours or more in a random sample of 50 hard drives is 0.0. This is, there is virtually no chance that this should happen. But this is what we found. Therefore, we conclude that the mean lifetime of the hard drives must have changed and it must be larger than the previous mean of 3250 hours. This type of reasoning will be placed in a more structured setting in the section on hypothesis testing.

2.23 $6 \pm 2.797\left(\dfrac{0.6}{\sqrt{25}}\right)$ or (5.66, 6.34) $\quad$ t(0.005, 24) = 2.797

2.25
```
Variable    N      Mean    StDev    SE Mean        95% CI
HWYMPG     147   28.1497   6.5337   0.5389    (27.0846, 29.2147)
```

2.27 If a null hypothesis is rejected at the 5% level of significance, it would also be rejected at the 10% level of significance because the 10% level requires less "evidence" to reject the hypothesis than does the 5% level. In other words, a test statistic that is more extreme than the 5% critical value is also more extreme than the 10% critical value.

2.29 Decision Rule: Reject H_0 if $t > 1.753$ $\quad$ t(0.05, 15) = 1.753
Do not reject H_0 if $t \leq 1.753$

Test Statistic: $t = \dfrac{20.5 - 20}{4/\sqrt{16}} = 0.5$

Decision: Do not reject H_0; standards are being met.

2.31 Test of mu = -18 vs > -18

```
                                                95%
                                              Lower
Variable    N      Mean    StDev   SE Mean   Bound      T      P
RET1YR     83   -13.2422  17.6712  1.9397   -16.4691   2.45  0.008
```

$H_0: \mu \leq -18.0$
$H_a: \mu > -18.0$

Decision Rule: Reject H_0 if $t > 1.645$ $t(0.05, 82) \approx 1.645$ (z value)
                Do not reject H_0 if $t \leq 1.645$

Test Statistic: $t = 2.45$

OR

Decision Rule: Reject H_0 if p value < 0.05
                Do not reject H_0 if p value ≥ 0.05

Test Statistic: p value $= 0.008$

Decision: Reject H_0; there is evidence that the population average return is greater than that of the S&P 500 index.

2.33 $s^2 = \dfrac{9(1.5)^2 + 9(1.0)^2}{10+10-2} = 1.625$ $t(0.01, 18) = 2.552$

$(15 - 11) \pm 2.552 \sqrt{1.625\left(\dfrac{1}{10} + \dfrac{1}{10}\right)}$ or $(2.55, 5.45)$

2.35 Because no information concerning the population variances is given, we assume that the population variances are not equal.

```
Two-sample T for ONEYRRET

LOAD   N    Mean   StDev   SE Mean
0     32   -10.1   23.5     4.2
1     51   -15.2   12.7     1.8

Difference = mu (0) - mu (1)
Estimate for difference:  5.06287
95% CI for difference:  (-4.04652, 14.17225)
T-Test of difference=0 (vs not =): T-Value= 1.12  P-Value= 0.268  DF= 42
```

The 95% confidence interval determined by MINITAB: (-4.04652, 14.17225)

2.37 Decision Rule: Reject H_0 if $t > 1.96$ or $t < -1.96$
Do not reject H_0 if $-1.96 \leq t \leq 1.96$

Test Statistic: $t = \dfrac{545 - 510}{\sqrt{\dfrac{104^2}{50} + \dfrac{95^2}{50}}} = 1.76$

Decision: Do not reject H_0; there is no difference in the population average test scores.

$z(0.025) = 1.96$ is used because df is large. Here we avoid computing Δ by reasoning that $\Delta \geq$ minimum $(n_1 - 1$ and $n_2 - 1)$ which means $\Delta \geq 49$ in this case, so z values can be used.

2.39 $H_0: \mu_1 - \mu_2 = 0$
$H_a: \mu_1 - \mu_2 \neq 0$

Decision Rule: Reject H_0 if $t > 2.048$ or $t < -2.048$ $t(0.025, 28) = 2.048$
Do not reject H_0 if $-2.048 \leq t \leq 2.048$

Test Statistic: $t = \dfrac{82 - 78}{\sqrt{7.625\left(\dfrac{1}{15} + \dfrac{1}{15}\right)}} = 3.97$

Decision: Reject H_0; there is a difference in mean rating scores for the two divisions.

2.41 Because no information concerning the population variances is given, we assume that the population variances are not equal.

```
Two-sample T for GRADRATE

PRIVATE   N    Mean    StDev   SE Mean
0         73   0.583   0.155   0.018
1         69   0.742   0.149   0.018


Difference = mu (0) - mu (1)
Estimate for difference:  -0.159015
95% upper bound for difference:  -0.116778
T-Test of difference=0 (vs <): T-Value= -6.23  P-Value= 0.000  DF= 139
```

Decision Rule: Reject H_0 if $t < -1.645$ $t(0.05, 139) \approx 1.645$ (z value)
Do not reject H_0 if $t \geq -1.645$

Test Statistic: $t = -6.23$

OR

Decision Rule: Reject H_0 if p value < 0.05
Do not reject H_0 if p value $\geq$ 0.05

Test Statistic: p value = 0.000

Decision: Reject H_0; there is evidence that the average public school graduation rate is less than the average for private schools.

2.43

```
Two-sample T for marketing vs finance

                                    SE
           N    Mean    StDev     Mean
marketing  12   1929     273       79
finance    12   1919     262       76


Difference = mu (marketing) - mu (finance)
Estimate for difference:   10.0000
95% CI for difference:   (-217.2583, 237.2583)
T-Test of difference=0(vs not =): T-Value=0.09 P-Value=0.928 DF=21
```

a $H_0: \mu_{Mark} - \mu_{Fina} = 0$
$H_a: \mu_{Mark} - \mu_{Fina} \neq 0$

Because no information concerning the population variances is given, we assume that the population variances are not equal.

Decision Rule: Reject H_0 if t > 2.080 or t < -2.080
Do not reject H_0 if -2.080 $\leq t \leq$ 2.080

Test Statistic: t = 0.09

OR

Decision Rule: Reject H_0 if p value < 0.05
Do not reject H_0 if p value $\geq$ 0.05

Test Statistic: p value = 0.928

b There is no difference in the average salaries.

2.45 Because no information concerning the population variances is given, we assume that the population variances are not equal.

```
Two-sample T for SALARY

MALE    N    Mean   StDev   SE Mean
0      61    5139    540      69
1      32    5957    691     122

Difference = mu (0) - mu (1)
Estimate for difference:   -818.023
95% upper bound for difference:  -582.958
T-Test of difference=0 (vs <): T-Value=-5.83  P-Value=0.000  DF = 51
```

a $H_0: \mu_0 - \mu_1 \geq 0$
   $H_a: \mu_0 - \mu_1 < 0$

where μ_0 = population average salary for females and μ_1 = population average salary for males.

Decision Rule: Reject H_0 if $t < -1.645$ $t(0.05, 51) = 1.645$ (z value)
Do not reject H_0 if $t \geq -1.645$

Test Statistic: $t = -5.83$
OR

Decision Rule: Reject H_0 if p value < 0.05
Do not reject H_0 if p value ≥ 0.05

Test Statistic: p value = 0.000

Decision: Reject H_0

Conclusion: Based on the result of the test, we conclude that the average starting salary for females is less than the average for males.

b Statistical evidence alone may not be sufficient to prove discrimination. Often, it must also be shown that there was intent to discriminate. However, when Harris Bank recognizes that a possible discriminatory situation exists, it should be concerned about correcting that situation.

c Later in the book, certain other variables are introduced into this problem; for example, education of the employee and years of experience. Variables such as these might have some sort of moderating effect on the result of the test reported in this problem and should be considered.

2.47   a   18.2%

   b   25.0%

   c   No, because 130,000 is not a limit of one of the classes. You could only approximate this percentage.

   d   Between 80,000 and 89,999 (because the median represents the 50th percentile).

2.49   Based on the time series plot, management may want to reconsider their decision. The time series plot shows a pattern of increasing errors over time which may indicate that the machine is experiencing wear and needs some type of maintenance. If the pattern continues, the subsequently drilled holes will not be of the correct diameter. Exercises 2.48 and 2.49 indicate the importance of using more than one type of graph, if appropriate, to examine data.

2.51

x	p(x)	xp(x)
0	0.90	0.0
1000	0.02	20.0
2000	0.04	80.0
3000	0.04	120.0
		220.0

$E(X) = 220$, so they should charge $220 per policy to break even.

2.53

x	p(x)	xp(x)
0	0.55	0.00
1	0.15	0.15
2	0.10	0.20
3	0.10	0.30
4	0.05	0.20
5	0.05	0.25
		1.10

   a $E(X) = 1.10$

   b $P(X > 2) = 0.10 + 0.05 + 0.05 = 0.20$

   c $P(X \le 4) = 1 - P(X = 5) = 1 - 0.05 = 0.95$

   d $P(2 < X < 5) = 0.10 + 0.05 = 0.15$

   e   0 (highest probability)

2.55 $x = 1000 - 1.29(25) = 967.75$ (About 968)

2.57 $P(X > 150) = P\left(Z > \dfrac{150 - 125}{37.50}\right) = P(Z > 0.67) = 0.5 - 0.2486 = 0.2514$

2.59 $P(-2 \leq \overline{X} - \mu \leq 2) = P\left(\dfrac{-2}{10/\sqrt{50}} \leq Z \leq \dfrac{2}{10/\sqrt{50}}\right) = P(-1.41 \leq Z \leq 1.41)$

$= 0.4207 + 0.4207 = 0.8414$

2.61 The interval (3.2, 4.5) is a 95% confidence interval estimate of the average sugar content. We are 95% confident that the **average** sugar content is between 3.2 and 4.5. The interval does not give us a range for 95% of the **individual** package contents however. This is not a correct interpretation of the interval.

CHAPTER THREE
Simple Regression Analysis

3.1 b

x	y	xy	x^2
5	12	60	25
6	11.5	69	36
7	14	98	49
8	15	120	64
9	15.4	138.6	81
10	15.3	153	100
11	17.5	192.5	121
56	100.7	831.1	476

$$b_1 = \frac{\sum xy - \frac{1}{n}\sum x \sum y}{\sum x^2 - \frac{1}{n}(\sum x)^2} = \frac{831.1 - \frac{1}{7}(56)(100.7)}{476 - \frac{1}{7}(56)^2} = 0.9107$$

$$b_0 = \bar{y} - b_1\bar{x} = \left(\frac{100.7}{7}\right) - 0.9107\left(\frac{56}{7}\right) = 7.1001$$

3.3 a

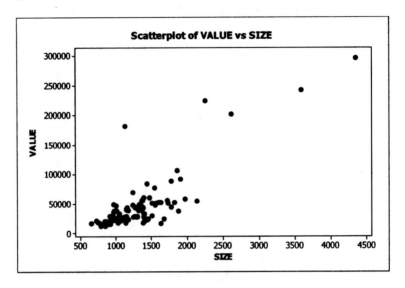

The regression equation is
VALUE = - 50035 + 72.8 SIZE

```
Predictor      Coef   SE Coef       T      P
Constant     -50035      7423   -6.74  0.000
SIZE         72.820     5.225   13.94  0.000
```

S = 27270.3   R-Sq = 66.5%   R-Sq(adj) = 66.1%

Analysis of Variance

```
Source          DF            SS            MS        F      P
Regression       1    1.44459E+11   1.44459E+11   194.25  0.000
Residual Error  98    72879341312     743666748
Total           99    2.17338E+11
```

Unusual Observations

```
Obs   SIZE    VALUE      Fit   SE Fit   Residual   St Resid
 76   2251   224182   113884     5571     110298       4.13R
 77   1126   182012    31961     2912     150051       5.53R
 78   2617   201597   140536     7298      61061       2.32RX
 97   3581   242690   210735    12117      31955       1.31 X
 98   4343   296251   266224    16022      30027       1.36 X
```

R denotes an observation with a large standardized residual.
X denotes an observation whose X value gives it large influence.

b

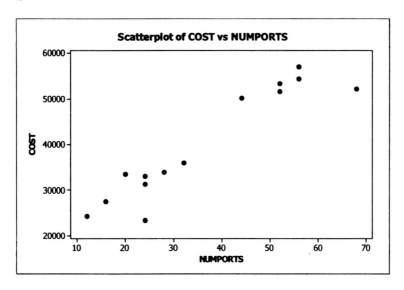

The regression equation is
COST = 16594 + 650 NUMPORTS

Predictor   Coef   SE Coef   T   P
Constant   16594   2687   6.18   0.000
NUMPORTS   650.17   66.91   9.72   0.000

S = 4306.91   R-Sq = 88.7%   R-Sq(adj) = 87.8%

Analysis of Variance

Source   DF   SS   MS   F   P
Regression   1   1751268376   1751268376   94.41   0.000
Residual Error   12   222594146   18549512
Total   13   1973862522

Unusual Observations

Obs   NUMPORTS   COST   Fit   SE Fit   Residual   St Resid
  1   68.0   52388   60805   2414   -8417   -2.36R
 10   24.0   23444   32198   1414   -8754   -2.15R

R denotes an observation with a large standardized residual.

c

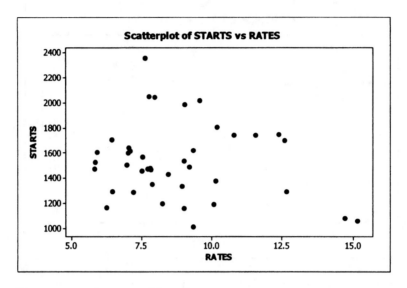

```
The regression equation is
STARTS = 1726 - 22.2 RATES


Predictor      Coef    SE Coef       T       P
Constant     1726.0      188.3    9.17   0.000
RATES        -22.23      20.75   -1.07   0.291


S = 297.173    R-Sq = 2.9%    R-Sq(adj) = 0.4%


Analysis of Variance

Source           DF        SS       MS      F      P
Regression        1    101430   101430   1.15  0.291
Residual Error   38   3355858    88312
Total            39   3457289


Unusual Observations

Obs   RATES   STARTS      Fit   SE Fit   Residual   St Resid
 10     7.6   2356.6   1557.1     53.1      799.5      2.73R
 19    14.7   1084.2   1399.2    131.3     -315.0     -1.18 X
 20    15.1   1062.2   1389.4    139.9     -327.2     -1.25 X

R denotes an observation with a large standardized residual.
X denotes an observation whose X value gives it large influence.
```

3.5 a Decision Rule: Reject H_0 if $t > 2.306$ or $t < -2.306$
Do not reject H_0 if $-2.306 \leq t \leq 2.306$

$t(0.025, 8) = 2.306$

Test Statistic: $t = \dfrac{b_1}{s_{b_1}} = \dfrac{2.0206}{0.0482} = 41.92$

Decision: Reject H_0

$$s_{b_1} = s_e \sqrt{\dfrac{1}{(n-1)s_x^2}} = 2.81\sqrt{\dfrac{1}{3400}} = 0.0482$$

$$(n-1)s_x^2 = \sum x^2 - \left(\dfrac{1}{n}\right)\left(\sum x\right)^2 = 39400 - \dfrac{1}{10}(600)^2 = 3400$$

x	$\hat{y}$	y	$y - \hat{y}$	$(y-\hat{y})^2$
40	78.9887	83	4.0113	16.0905
30	58.7827	60	1.2173	1.4818
70	139.6067	138	-1.6067	2.5815
90	180.0187	180	-0.0187	0.0003
50	99.1947	97	-2.1947	4.8167
60	119.4007	118	-1.4007	1.9620
70	139.6067	140	0.3933	0.1547
40	78.9887	75	-3.9887	15.9097
80	159.8127	159	-0.8127	0.6605
70	139.6067	144	4.3933	19.3011
				62.9588

$$s_e = \sqrt{\dfrac{\sum(y_i - \hat{y}_i)^2}{n-2}} = \sqrt{\dfrac{62.9588}{8}} = \sqrt{7.87} = 2.81$$

(Note: there are other formulas to compute s_e).

b Yes. Hours of labor and number of items produced appear to be linearly related.

c   Decision Rule: Reject H_0 if $t > 2.306$ or $t < -2.306$
Do not reject H_0 if $-2.306 \leq t \leq 2.306$

$t(0.025, 8) = 2.306$

Test Statistic: $t = \dfrac{b_0}{s_{b_0}} = \dfrac{-1.8353}{3.025} = -0.61$

Decision: Do not reject H_0

$$s_{b_0} = s_e\sqrt{\dfrac{1}{n} + \dfrac{\bar{x}^2}{(n-1)s_x^2}} = 2.81\sqrt{\dfrac{1}{10} + \dfrac{(60)^2}{3400}} = 3.025$$

d   The intercept of the line representing the relationship between number of items and labor is not significantly different from 0. (Note: This test does not suggest anything about whether there is or is not a relationship between number of items and labor.)

3.7 a   Hypotheses: H_0: $\beta_1 = 0$
H_a: $\beta_1 \neq 0$

Decision Rule: Reject H_0 if $t > 2.101$ or $t < -2.101$
Do not reject H_0 if $-2.101 \leq t \leq 2.101$

$t(0.025, 18) = 2.101$

Test Statistic: $t = 10.70$

OR

Decision Rule: Reject H_0 if p value < 0.05
Do not reject H_0 if p value ≥ 0.05

Test Statistic: p value $= 0.000$

Decision: Reject H_0

b   Sales and advertising appear to be linearly related. The relationship is direct, suggesting that increases in advertising expenditures may result in an increase in sales. (Caution: see Section 3.7 in text. There may not be a causal relationship here.)

c $\hat{y} = -57.281 + 17.57x$ or

SALES = -57.281 + 17.57 ADV

d Our prediction of SALES would increase by 17.57(10) = 175.7 or $17,570 (SALES and ADV are expressed in $100)

e $-57.281 \pm 1.734(509.75)$ or $\quad$ (-941.19, 826.63)

f $17.570 \pm 2.101(1.642)$ or $\quad$ (14.12, 21.02)

g Decision Rule: Reject H_0 if $t > 2.101$ or $t < -2.101$
Do not reject H_0 if $-2.101 \leq t \leq 2.101$

$t(0.025, 18) = 2.101$

Test Statistic: $t = \dfrac{b_1 - \beta_1^*}{s_{b_1}} = \dfrac{17.57 - 20}{1.642} = -1.48$

Decision: Do not reject H_0

h The slope of the regression line is not significantly different from 20.

3.9 a $R^2 = \dfrac{SSR}{SST} = \dfrac{13881.4}{13944.4} = 0.9955$

$SST = \sum y^2 - \left(\dfrac{1}{n}\right)\left(\sum y\right)^2 = 156508 - \left(\dfrac{1}{10}\right)(1194)^2 = 13944.4$

SSR = SST − SSE = 13944.4 − 62.9588 = 13881.44 (see problem 3.5 for computation of SSE).

b 99.55%

c Decision Rule: Reject H_0 if $F > 5.32$
Do not reject H_0 if $F \leq 5.32$

$F(0.05; 1,8) = 5.32$

Test Statistic: $F = \dfrac{MSR}{MSE} = \dfrac{13881.44}{62.9588/8} = 1763.88$

Decision: Reject H₀

d Yes. Hours of labor and number of items produced appear to be linearly related.

3.11 a 86.4%

b Decision Rule: Reject H_0 if $F > 4.41$
Do not reject H_0 if $F \leq 4.41$

$F(0.05; 1,18) = 4.41$

Test Statistic: $F = 114.54$

OR

Decision Rule: Reject H_0 if p value < 0.05
Do not reject H_0 if p value ≥ 0.05

Test Statistic: p value $= 0.000$

Decision: Reject H_0

3.13 a $\hat{y}_m = -1.8353 + 2.0206(60) = 119.4007$

b $\hat{y}_m \pm t(\alpha/2, n-2) s_m$

$$s_m^2 = s_e^2 \left(\frac{1}{n} + \frac{(x_m - \bar{x})^2}{(n-1)s_x^2} \right) = (7.87) \left(\frac{1}{10} + \frac{(60-60)^2}{3400} \right) = 0.787$$

$s_m = \sqrt{s_m^2} = \sqrt{0.787} = 0.8871$

$119.4007 \pm 2.306(0.8871)$ or $(117.36, 121.45)$

c $\hat{y}_p = -1.8353 + 2.0206(60) = 119.4007$

d $\hat{y}_p \pm t(\alpha/2, n-2) s_p$

$$s_p^2 = s_e^2\left(1 + \frac{1}{n} + \frac{(x_p - \bar{x})^2}{(n-1)s_x^2}\right) = (7.87)\left(1 + \frac{1}{10} + \frac{(60-60)^2}{3400}\right) = 8.66$$

$$s_p = \sqrt{s_p^2} = \sqrt{8.66} = 2.94$$

119.4007 ± 2.306(2.94) or   (112.62, 126.18)

3.15 a $\hat{y}_m$ = 4335 (in $100)

95% confidence interval estimate: (4007, 4663) (in $100)

b

x	Point Estimate 95%	Prediction Interval
20000	3457	2131, 4783
25000	4335	3043, 5627
30000	5214	3933, 6494
35000	6092	4800, 7384

Note: Both point estimates and interval limits are in $100.

3.17 a $\hat{y}_m$ = -57 + 17.6(600) = 10,503 ($1,050,300)

Yes. The regression equation was developed using a range of values for the x variable (ADV) of 160 to 415. The value 600 is well outside this range. Caution should be exercised if this estimate is used, because the relationship between SALES and ADV has been observed only over the range 160 to 415. It is not known whether the same relationship will serve as well outside this range.

b   Disagree. The model was developed over the range of 160 to 415 for the x variable (ADV). Because ADV = 0 is outside this range, the resulting forecasts cannot be depended upon to make sense. Still, the least squares method must choose a value as a y intercept. In this case, the intercept value that minimized the error sum of squares was -5700.

3.19 a   Decision Rule: Reject H_0 if $t > 2.074$ or $t < -2.074$
Do not reject H_0 if $-2.074 \le t \le 2.074$

t(0.025, 22) = 2.074

Test Statistic: $t = \dfrac{b_1}{s_{b_1}} = \dfrac{2.0}{0.25} = 8.0$

Decision: Reject H_0

b Decision Rule: Reject H_0 if $F > 2.95$
Do not reject H_0 if $F \leq 2.95$

$F(0.1; 1, 22) = 2.95$

Test Statistic: $F = \dfrac{MSR}{MSE} = \dfrac{SSR/1}{SSE/(n-2)} = \dfrac{87.273}{30/22} = 64$

$SSR = SST - SSE = 117.273 - 30 = 87.273$

Decision: Reject H_0

c $R^2 = \dfrac{SSR}{SST} = \dfrac{87.273}{117.273} = 0.744$ or 74.4%

3.21

Source	DF	SS	MS	F
Regression	1	1000	1000	100
Error (Residual)	80	800	10	
Total	81	1800		

3.23 a Hypotheses: $H_0: \beta_1 = 0$
$H_a: \beta_1 \neq 0$

Decision Rule: Reject H_0 if $t > 1.645$ or $t < -1.645$
Do not reject H_0 if $-1.645 \leq t \leq 1.645$

$t(0.05, 91) \approx 1.645$

Test Statistic: $t = 4.31$

OR

Decision Rule: Reject H_0 if p value < 0.1
Do not reject H_0 if p value ≥ 0.1

Test Statistic: p value $= 0.000$

OR

Decision Rule: Reject H_0 if $F \geq 2.79$
　　　　　　　　 Do not reject H_0 if $F < 2.79$

$F(0.10; 1, 91) \approx 2.79$

Test Statistic: $F = 18.6$

Decision: Reject H_0

Conclusion: There is a linear relationship between salary and education.

b   17%

c $\hat{y}_p = 3818.6 + 128.1(12) = 5355.8$

d $\hat{y}_m = 3818.6 + 128.1(12) = 5355.8$

e   As will be seen when this problem is continued later in this text, other factors might include experience and, unfortunately in this case, whether the employee is male or female. You can probably think of other factors that might be useful.

3.25 The MINITAB output for the regression of consumption on income follows. This output will be used to help answer the questions.

```
The regression equation is
CONS = 2521 + 0.827 INCOME


Predictor        Coef    SE Coef        T        P
Constant         2521       2620     0.96    0.358
INCOME        0.82690    0.07110    11.63    0.000


S = 4387.72    R-Sq = 93.1%    R-Sq(adj) = 92.4%

Analysis of Variance

Source           DF           SS           MS         F        P
Regression        1   2603912710   2603912710    135.25    0.000
Residual Error   10    192521012     19252101
Total            11   2796433722


Unusual Observations

Obs  INCOME    CONS     Fit   SE Fit  Residual  St Resid
  5   56000   40176   48828     2111     -8652    -2.25R

R denotes an observation with a large standardized residual.
```

a $\hat{y} = 2521 + 0.827x$ or   CONS = 2521 + 0.827 INCOME

b 93.1%

c $0.82690 \pm (1.812)(0.0711)$ or   (0.70, 0.96)

d Decision Rule: Reject H_0 if $t > 2.228$ or $t < -2.228$
Do not reject H_0 if $-2.228 \leq t \leq 2.228$

t(0.025, 10) = 2.228

Test Statistic: $t = 11.63$

OR

Decision Rule: Reject H_0 if p value < 0.05
Do not reject H_0 if p value ≥ 0.05

e	Test Statistic:	p value = 0.000
	Decision:	Reject H_0
	Decision Rule:	Reject H_0 if $F > 4.96$ Do not reject H_0 if $F \leq 4.96$

$F(0.05; 1, 10) = 4.96$

Test Statistic: $F = 135.25$

OR

Decision Rule: Reject H_0 if p value < 0.05
Do not reject H_0 if p value ≥ 0.05

Test Statistic: p value = 0.000

Decision: Reject H_0

f   No. The F test can be used only to test the two-tailed hypotheses.

g   Decision Rule: Reject H_0 if $t > 2.228$ or $t < -2.228$
Do not reject H_0 if $-2.228 \leq t \leq 2.228$

$t(0.025, 10) = 2.228$

Test Statistic: $t = \dfrac{b_1 - \beta_1^*}{s_{b_1}} = \dfrac{0.82690 - 1}{0.0711} = -2.43$

Decision: Reject H_0

Conclusion: The coefficient of INCOME is significantly different from 1.

3.27 The MINITAB output for the regression of new construction on a linear trend variable follows. This output will be used to help answer the questions.

```
The regression equation is
NEWCON = 368 + 43.0 TREND


Predictor     Coef    SE Coef      T        P
Constant    368.23     10.19     36.13    0.000
TREND        42.993     1.503    28.61    0.000


S = 15.7599    R-Sq = 98.9%    R-Sq(adj) = 98.8%


Analysis of Variance

Source            DF      SS        MS         F        P
Regression         1   203321    203321    818.61    0.000
Residual Error     9     2235       248
Total             10   205557
```

a $\hat{y} = 368 + 43t$ or   NEWCON = 368 + 43TREND

b   98.9%

c   Using the R-square value, the equation fits the observed data very well. However this is no guarantee that the equation will predict new construction accurately in future years. This depends on whether the same linear increase in new construction continues into the future.

d The following MINITAB output was used to obtain the desired prediction and prediction interval.

```
The regression equation is
NEWCON = 368 + 43.0 TREND


Predictor      Coef    SE Coef       T       P
Constant     368.23      10.19   36.13   0.000
TREND        42.993      1.503   28.61   0.000


S = 15.7599    R-Sq = 98.9%    R-Sq(adj) = 98.8%


Analysis of Variance

Source          DF      SS       MS       F        P
Regression       1   203321   203321   818.61    0.000
Residual Error   9     2235      248
Total           10   205557


Predicted Values for New Observations

New
Obs     Fit   SE Fit        95% CI              95% PI
 1   884.14    10.19   (861.08, 907.19)    (841.68, 926.59)
 2   927.13    11.54   (901.02, 953.24)    (882.94, 971.32)


Values of Predictors for New Observations

New
Obs   TREND
 1    12.0
 2    13.0
```

Point prediction of *y* for 2002: 884.14
95% prediction interval of *y* for 2002: (841.68, 926.59)

Point prediction of *y* for 2003: 927.13
95% prediction interval of *y* for 2003: (882.94, 971.32)

e See part c.

3.29 The MINITAB output for the regression of wheat shipments on exchange rates follows. This output will be used to help answer the questions.

```
The regression equation is
SHIPMENT = 1969 + 7.86 EXCHRATE


Predictor      Coef    SE Coef       T        P
Constant     1969.1      413.6    4.76    0.000
EXCHRATE      7.862      3.760    2.09    0.038


S = 819.383    R-Sq = 3.2%    R-Sq(adj) = 2.5%


Analysis of Variance

Source            DF         SS         MS       F       P
Regression         1    2935648    2935648    4.37   0.038
Residual Error   133   89294612     671388
Total            134   92230260


Unusual Observations

Obs   EXCHRATE   SHIPMENT      Fit   SE Fit   Residual   St Resid
 93       107      5284.0   2812.3     70.7     2471.7      3.03R
129       151      6605.0   3156.8    175.3     3448.2      4.31RX
130       154      3736.0   3178.2    184.7      557.8      0.70 X
131       151      2648.0   3158.9    176.2     -510.9     -0.64 X
132       156      3591.0   3195.0    192.1      396.0      0.50 X
133       160      1897.0   3229.0    207.4    -1332.0     -1.68 X
134       166      2327.0   3274.2    227.8     -947.2     -1.20 X
135       166      1576.0   3273.7    227.5    -1697.7     -2.16RX


R denotes an observation with a large standardized residual.
X denotes an observation whose X value gives it large influence.
```

a $\hat{y} = 1969 + 7.86x$ or

SHIPMENT = 1969 + 7.86 EXCHRATE

b   Hypotheses: $H_0: \beta_1 = 0$
$H_a: \beta_1 \neq 0$

Decision Rule: Reject H_0 if $t > 1.96$ or $t < -1.96$
Do not reject H_0 if $-1.96 \leq t \leq 1.96$

$t(0.025, 133) \approx 1.96$ (z value)

Test Statistic: $t = 2.09$

OR

Decision Rule: Reject H_0 if p value < 0.05
Do not reject H_0 if p value ≥ 0.05

Test Statistic: p value $= 0.038$

OR

Decision Rule: Reject H_0 if $F > 3.84$
Do not reject H_0 if $F \leq 3.84$

$F(0.05; 1, 133) \approx 3.84$ (or 3.92)

Test Statistic: $F = 4.37$

Decision: Reject H_0; wheat shipments and exchange rates are linearly related.

c 3.2%

d 7.862 ± 1.96 (3.760) or (0.49, 15.23)

3.31 The MINITAB output for the regression of each company's return on the market return follows. This output will be used to help answer the question.

Regression of WALMARTRET on MKTRET

```
The regression equation is
WALMARTRET = 0.0198 + 0.732 MKTRET


Predictor       Coef    SE Coef      T        P
Constant     0.01977    0.01082   1.83    0.073
MKTRET        0.7317     0.1912   3.83    0.000


S = 0.0838394    R-Sq = 20.2%    R-Sq(adj) = 18.8%


Analysis of Variance

Source           DF        SS         MS        F        P
Regression        1   0.10292    0.10292    14.64    0.000
Residual Error   58   0.40768    0.00703
Total            59   0.51060


Unusual Observations

Obs   MKTRET   WALMARTRET     Fit    SE Fit   Residual   St Resid
  8   -0.158      -0.0654  -0.0956   0.0322     0.0302      0.39 X
 10    0.074       0.2643   0.0742   0.0177     0.1901      2.32R
 25   -0.040      -0.2080  -0.0092   0.0133    -0.1987     -2.40R
 32    0.076      -0.1365   0.0757   0.0180    -0.2122     -2.59R
 35   -0.103       0.1501  -0.0556   0.0226     0.2058      2.55R
```

R denotes an observation with a large standardized residual.
X denotes an observation whose X value gives it large influence.

Regression of DELLRET on MKTRET

The regression equation is
DELLRET = 0.0280 + 1.67 MKTRET

```
Predictor        Coef    SE Coef       T      P
Constant      0.02800    0.01846    1.52  0.135
MKTRET         1.6679     0.3260    5.12  0.000
```

S = 0.142958   R-Sq = 31.1%   R-Sq(adj) = 29.9%

Analysis of Variance

```
Source          DF        SS       MS       F      P
Regression       1   0.53482  0.53482   26.17  0.000
Residual Error  58   1.18534  0.02044
Total           59   1.72016
```

Unusual Observations

```
Obs  MKTRET  DELLRET      Fit   SE Fit  Residual  St Resid
  8  -0.158  -0.0791  -0.2350   0.0549    0.1558      1.18 X
 37   0.040   0.4982   0.0940   0.0224    0.4042      2.86R
```

R denotes an observation with a large standardized residual.
X denotes an observation whose X value gives it large influence.

Regression of SABRERET on MKTRET

```
The regression equation is
SABRERET = 0.0025 + 1.47 MKTRET


Predictor        Coef     SE Coef        T        P
Constant      0.00252     0.01390     0.18    0.857
MKTRET         1.4706      0.2455     5.99    0.000


S = 0.107652    R-Sq = 38.2%    R-Sq(adj) = 37.2%

Analysis of Variance

Source             DF         SS         MS        F        P
Regression          1    0.41578    0.41578    35.88    0.000
Residual Error     58    0.67216    0.01159
Total              59    1.08794


Unusual Observations

Obs   MKTRET   SABRERET      Fit   SE Fit  Residual  St Resid
  8   -0.158    -0.2062  -0.2293   0.0413    0.0231      0.23 X
 35   -0.103     0.0673  -0.1490   0.0291    0.2163      2.09R
 45   -0.091    -0.3661  -0.1320   0.0266   -0.2341     -2.24R

R denotes an observation with a large standardized residual.
X denotes an observation whose X value gives it large influence.
```

a Wal-Mart: 0.732
Dell: 1.67
Sabre: 1.47

b For each company we want to test the hypotheses:

Hypotheses: $H_0: \beta_1 = 0$
$H_a: \beta_1 \neq 0$

Using the *p* value we have the following decision rule:

Decision Rule: Reject H_0 if *p* value < 0.05
Do not reject H_0 if *p* value $\geq$ 0.05

Test Statistic: The *p* values for each of the companies are each 0.000

Decision: Reject H_0; each company's returns and the market returns are linearly related.

c   For each company we want to test the hypotheses:

Hypotheses: $H_0: \beta_1 = 1.0$
$H_a: \beta_1 \neq 1.0$

Decision Rule: Reject H_0 if $t > 1.96$ or $t < -1.96$
Do not reject H_0 if $-1.96 \leq t \leq 1.96$

$t(0.025, 58) \approx 1.96$ (z value)

Test Statistic: Wal-Mart: $t = \dfrac{0.7317 - 1.0}{0.1912} = -1.40$

Dell: $t = \dfrac{1.6679 - 1.0}{0.3260} = 2.05$

Sabre: $t = \dfrac{1.4706 - 1.0}{0.2455} = 1.92$

Decision: Wal-Mart: Do not reject H_0

Dell: Reject H_0

Sabre: Do not reject H_0

Conclusion: We cannot reject the hypothesis that the beta is one for Wal-Mart and Sabre, but it is rejected for Dell.

d   Hypotheses: $H_0: \beta_1 \leq 1.0$
$H_a: \beta_1 > 1.0$

Decision Rule: Reject H_0 if $t > 1.645$
Do not reject H_0 if $t \leq 1.645$

$t(0.05, 58) \approx 1.645$ (z value)

Test Statistic: $t = \dfrac{1.6679 - 1.0}{0.3260} = 2.05$

Decision: Reject H_0; Dell's beta coefficient is greater than one.

e   Hypotheses: $H_0: \beta_1 \geq 1.0$
                    $H_a: \beta_1 < 1.0$

Decision Rule: Reject H_0 if $t < -1.645$
Do not reject H_0 if $t \geq -1.645$

$t(0.05, 58) \approx 1.645$ (z value)

Test Statistic: $t = \dfrac{0.7317 - 1.0}{0.1912} = -1.40$

Decision: Do not reject H_0; Wal-Mart's beta coefficient is not less than one.

3.33 The time-series plot of work orders is shown below. Obviously, the number of open work orders is decreasing, as desired. Our goal is to predict the time when the number of open work orders will reach 1000. To do this, a trend line is fit to the work order data. Then the equation for the trend line will be solved to determine the number of days required to reach the 1000 work order level.

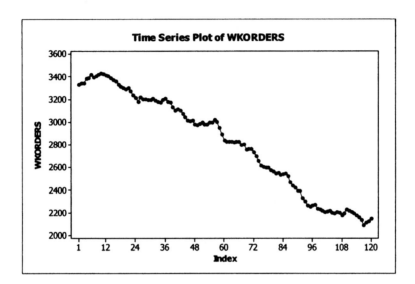

The regression of work orders on the trend variable results in the following equation:

WKORDER = 3559 − 12.3 TREND

We want to know the value of the trend variable when WKORDER = 1000, so we solve the following equation:

1000 = 3559 − 12.3 TREND

The solution is $TREND = \dfrac{(1000 - 3559)}{-12.3} = 208$.

So, our prediction for the time when work orders reaches 1000 would be day 208. Because the number of days included in the sample is 120, the prediction is 208−120 = 88 days from now.

The primary assumption here is that past behavior of the dependent variable, work orders, will be a good predictor of what will happen in the future. We assume that the downward trend in work orders will continue at the same rate that it has in the past. The trend should be monitored in coming days to see if this pattern persists. At this point, however, the prediction of 88 days to reach the desired goal seems to be a reasonable one.

Regression of WKORDERS on linear trend variable

```
The regression equation is
WKORDERS = 3559 - 12.3 TREND


Predictor        Coef    SE Coef         T        P
Constant      3558.50      12.45    285.91    0.000
TREND        -12.3370     0.1785    -69.10    0.000


S = 67.7445     R-Sq = 97.6%     R-Sq(adj) = 97.6%


Analysis of Variance

Source            DF         SS         MS         F        P
Regression         1   21915488   21915488   4775.32    0.000
Residual Error   118     541540       4589
Total            119   22457028


Unusual Observations

Obs   TREND   WKORDERS       Fit   SE Fit   Residual   St Resid
  1       1    3332.00   3546.17    12.29    -214.17     -3.21R
  2       2    3348.00   3533.83    12.14    -185.83     -2.79R
  3       3    3348.00   3521.49    11.98    -173.49     -2.60R
 56      56    3021.00   2867.63     6.24     153.37      2.27R
 57      57    3004.00   2855.30     6.22     148.70      2.20R

R denotes an observation with a large standardized residual.
```

3.35 Of the three possible explanatory variables, student faculty ratio provides the highest R-square. That regression follows:

Regression of GRADRATE4 on SFACRATIO

```
The regression equation is
GRADRATE4 = 1.06 - 0.0394 SFACRATIO


Predictor        Coef     SE Coef         T        P
Constant      1.06496     0.03989     26.70    0.000
SFACRATIO   -0.039372    0.002882    -13.66    0.000


S = 0.169233    R-Sq = 49.2%    R-Sq(adj) = 48.9%


Analysis of Variance

Source           DF        SS        MS         F        P
Regression        1    5.3440    5.3440    186.59    0.000
Residual Error  193    5.5275    0.0286
Total           194   10.8715


Unusual Observations

Obs   SFACRATIO   GRADRATE4      Fit   SE Fit   Residual   St Resid
  2        16.0      0.8100   0.4350   0.0146     0.3750       2.22R
 18        13.0      0.1200   0.5531   0.0121    -0.4331      -2.57R
 24        19.0      0.6900   0.3169   0.0207     0.3731       2.22R
 32        19.0      0.6700   0.3169   0.0207     0.3531       2.10R
 62        11.0      0.2200   0.6319   0.0137    -0.4119      -2.44R
 63        13.0      0.2100   0.5531   0.0121    -0.3431      -2.03R
 67        24.0      0.2500   0.1200   0.0334     0.1300       0.78 X
 80        11.0      0.2100   0.6319   0.0137    -0.4219      -2.50R
 89        14.0      0.1000   0.5137   0.0123    -0.4137      -2.45R
 91        15.0      0.0900   0.4744   0.0132    -0.3844      -2.28R
 96         3.0      0.7100   0.9468   0.0318    -0.2368      -1.42 X
182        13.0      0.0000   0.5531   0.0121    -0.5531      -3.28R
192        11.0      0.0000   0.6319   0.0137    -0.6319      -3.75R

R denotes an observation with a large standardized residual.
X denotes an observation whose X value gives it large influence.
```

Summary of Results for Other Explanatory Variables

Variable	p-value	R-square
ADMISRATE	0.000	40.7%
AVGDEBT	0.457	0.3%

3.37

Regression of ATTENDANCE on WINS

```
The regression equation is
ATTENDANCE = 122188 + 22521 WINS


Predictor      Coef    SE Coef      T      P
Constant     122188     649930   0.19  0.852
WINS          22521       8598   2.62  0.015


S = 521990    R-Sq = 20.9%    R-Sq(adj) = 17.8%

Analysis of Variance

Source           DF          SS            MS         F      P
Regression        1   1.86939E+12   1.86939E+12    6.86  0.015
Residual Error   26   7.08432E+12   2.72474E+11
Total            27   8.95371E+12


Unusual Observations

Obs   WINS   ATTENDANCE       Fit   SE Fit   Residual  St Resid
 10   38.0       565637    977989   330727    -412352     -1.02 X
 23   49.0      1845208   1225721   242102     619487      1.34 X
 28   67.0      2813854   1631100   118872    1182754      2.33R

R denotes an observation with a large standardized residual.
X denotes an observation whose X value gives it large influence.
```

a   ATTENDANCE = 122188 + 22521 WINS

b   The Cubs organization should be cautious when forecasting for 110 wins. The range of the *x* variable for this regression was from 38 to 96. We do not know if the same relationship will hold outside this range. Also, it is a relatively weak relationship (R-square = 20.9%) to begin with.

CHAPTER 4
Multiple Regression Analysis

4.1   **a** $\hat{y} = 51.72 + 0.95x_1 + 2.47x_2 + 0.05x_3 - 0.05x_4$

OR

COST = 51.72 + 0.95 PAPER + 2.47 MACHINE + 0.05 OVERHEAD − 0.05 LABOR

     b   Hypotheses: $H_0: \beta_1 = \beta_2 = \beta_3 = \beta_4 = 0$
                           H_a: At least one of the coefficients ($\beta_1, \beta_2, \beta_3, \beta_4$) is not equal to 0.

Decision Rule: Reject H_0 if $F > 2.82$ $F(0.05; 4, 22) = 2.82$
                            Do not reject H_0 if $F \leq 2.82$

Test Statistic: $F = 4629.17$

OR

Decision Rule: Reject H_0 if p value < 0.05
                           Do not reject H_0 if p value ≥ 0.05

Test Statistic: p value = 0.000

Decision: Reject H_0

Conclusion: At least one of the coefficients ($\beta_1, \beta_2, \beta_3, \beta_4$) is not equal to 0. In other words, at least one of the variables (x_1, x_2, x_3, x_4) is helping to explain a significant amount of the variation in y.

     c $b_2 = 2.47$ $2.47 \pm (2.074)(0.47)$

     d   Hypotheses: $H_0: \beta_1 = 1.0$
                           $H_a: \beta_1 \neq 1.0$

Decision Rule: Reject H_0 if $t > 2.074$ or $t < -2.074$
                            Do not reject H_0 if $-2.074 \leq t \leq 2.074$

$t(0.025, 22) = 2.074$

Test Statistic: $t = \dfrac{b_1 - \beta_1^*}{s_{b_1}} = \dfrac{0.95 - 1.0}{0.12} = -0.42$

Decision: Do not reject H_0

Conclusion: The true marginal cost of output associated with total production of paper is 1.

e   99.9%

f   99.9%

g   The regression equation can be used to identify factors that are related to cost. After doing this, these factors might be useful in reducing cost. For example, machine hours are related to cost. Obviously we cannot just start reducing machine hours, because these hours are part of the manufacturing process. But there may be a way to use the hours more efficiently, resulting in a reduction of overall hours without sacrificing production and a decrease in cost. The regression equation shows the influence of a one-unit reduction of the variables included on cost.

4.3   Hypotheses: H_0: $\beta_3 = \beta_4 = 0$
H_a: At least one of the coefficients (β_3, β_4) is not equal to zero.

Decision Rule: Reject H_0 if $F > 3.44$ $F(0.05; 2, 22) = 3.44$
Do not reject H_0 if $F \leq 3.44$

Test Statistic: $F = \dfrac{(2895 - 2699)/2}{2699/22} = 0.80$

Decision: Do not reject H_0

Conclusion: Both β_3 and β_4 are equal to 0. Neither OVERHEAD nor LABOR adds significantly to the model's ability to explain the variation in COST. Choose the REDUCED model.

4.5 **a** $\hat{y}_i = 0.16077 + 0.97774 y_{i-1}$

OR RATES$_i$ = 0.16077 + 0.97774 RATES$_{i-1}$

b Hypotheses: $H_0: \beta_1 = 0.0$
$H_a: \beta_1 \neq 0.0$

Decision Rule: Reject H_0 if $t > 1.96$ or $t < -1.96$
Do not reject H_0 if $-1.96 \leq t \leq 1.96$

$t(0.025, 213) \approx 1.96$ (z value)

Test Statistic: $t = 97.60$

OR

Decision Rule: Reject H_0 if p value < 0.05
Do not reject H_0 if p value ≥ 0.05

Test Statistic: p value = 0.000

Decision: Reject H_0

c 97.8%

d Using MINITAB, the forecast of the mortgage rate in January 2003 is 6.0761, as shown in the output on the next page. By hand, the forecast would be

RATES = 0.16077 + 0.97774 (6.05) = 6.0761

Forecasts for 2003 using MINITAB:

Date	Fit	SE Fit	95% C.I.	95% P.I.
1/03	6.0761	0.0310	(6.0151, 6.1371)	(5.6080, 6.5443)
2/03	6.1016	0.0307	(6.0410, 6.1622)	(5.6335, 6.5697)
3/03	6.1266	0.0305	(6.0664, 6.1867)	(5.6585, 6.5946)
4/03	6.1510	0.0303	(6.0913, 6.2107)	(5.6830, 6.6190)
5/03	6.1749	0.0301	(6.1155, 6.2342)	(5.7069, 6.6428)
6/03	6.1982	0.0299	(6.1393, 6.2571)	(5.7303, 6.6661)
7/03	6.2210	0.0297	(6.1625, 6.2795)	(5.7532, 6.6888)
8/03	6.2433	0.0295	(6.1851, 6.3015)	(5.7755, 6.7111)
9/03	6.2651	0.0293	(6.2073, 6.3229)	(5.7974, 6.7328)
10/03	6.2864	0.0291	(6.2290, 6.3438)	(5.8187, 6.7541)
11/03	6.3072	0.0290	(6.2502, 6.3643)	(5.8396, 6.7749)
12/03	6.3276	0.0288	(6.2708, 6.3843)	(5.8600, 6.7952)

Forecasts versus Actual Values:

Date	Forecast	Actual Value
1/03	6.0761	5.92
2/03	6.1016	5.84
3/03	6.1266	5.75
4/03	6.1510	5.81
5/03	6.1749	5.48
6/03	6.1982	5.23
7/03	6.2210	5.63
8/03	6.2433	6.26
9/03	6.2651	6.15
10/03	6.2864	5.95
11/03	6.3072	5.93
12/03	6.3276	5.91

Regression of Mortgage Rate on Lagged Dependent Variable with Forecast

```
The regression equation is
RATE = 0.161 + 0.978 RATESL1


215 cases used, 1 case contains missing values


Predictor      Coef    SE Coef       T       P
Constant    0.16077    0.08854    1.82   0.071
RATESL1     0.97774    0.01002   97.60   0.000


S = 0.235473    R-Sq = 97.8%    R-Sq(adj) = 97.8%


Analysis of Variance

Source              DF        SS       MS        F       P
Regression           1    528.13   528.13  9524.82   0.000
Residual Error     213     11.81     0.06
Total              214    539.94


Unusual Observations

Obs   RATESL1      RATE      Fit   SE Fit   Residual   St Resid
  2      13.1   12.9200  12.9496   0.0468    -0.0296      -0.13 X
  3      12.9   13.1700  12.7932   0.0453     0.3768       1.63 X
  4      13.2   13.2000  13.0376   0.0477     0.1624       0.70 X
  5      13.2   12.9100  13.0670   0.0479    -0.1570      -0.68 X
  6      12.9   12.2200  12.7834   0.0452    -0.5634      -2.44RX
 15      10.7   10.0800  10.6324   0.0258    -0.5524      -2.36R
 18      10.1   10.6800  10.0751   0.0216     0.6049       2.58R
 28       9.0    9.8300   8.9995   0.0164     0.8305       3.54R
 29       9.8   10.6000   9.7720   0.0197     0.8280       3.53R
```

```
33        10.3    10.8900    10.2608    0.0230     0.6292     2.68R
35        11.3    10.6500    11.1701    0.0303    -0.5201    -2.23R
38        10.4     9.8900    10.3586    0.0237    -0.4686    -2.00R
54        10.8    10.2000    10.6910    0.0263    -0.4910    -2.10R
111        7.2     7.6800     7.1516    0.0223     0.5284     2.25R
112        7.7     8.3200     7.6698    0.0190     0.6502     2.77R
135        7.1     7.6200     7.0832    0.0228     0.5368     2.29R

R denotes an observation with a large standardized residual.
X denotes an observation whose X value gives it large influence.

Predicted Values for New Observations

New
Obs      Fit   SE Fit        95% CI              95% PI
  1   6.0761   0.0310   (6.0151, 6.1371)   (5.6080, 6.5443)


Values of Predictors for New Observations

New
Obs   RATESL1
  1     6.05
```

One of the problems encountered when trying to produce forecasts for February or subsequent months is that the actual value in the previous month is not available at the end of 2002. One way around this problem is to use the forecast value for the previous month to generate the next month's forecast. To develop the 2/03 forecast, use 6.0761 as the value of the lagged variable. Then use the forecast for 2/03 as the value of the lagged variable to generate the forecast for 3/03, and so on. As forecasts are generated farther into the future, we expect them to be less accurate in part because we are using previous month's forecasts to develop forecasts for future months. But, until we know the true values, there is not much alternative.

e Hypotheses: $H_0: \beta_0 = 0.0$
                   $H_a: \beta_0 \neq 0.0$

Decision Rule: Reject H_0 if $t > 1.96$ or $t < -1.96$
                Do not reject H_0 if $-1.96 \leq t \leq 1.96$

                $t(0.025, 213) \approx 1.96$ (z value)

Test Statistic: $t = 1.82$

OR

Decision Rule: Reject H_0 if p value < 0.05
Do not reject H_0 if p value ≥ 0.05

Test Statistic: p value $= 0.071$

Decision: Do not reject H_0

f   Hypotheses: $H_0: \beta_1 = 1.0$
$H_a: \beta_1 \neq 1.0$

Decision Rule: Reject H_0 if $t > 1.96$ or $t < -1.96$
Do not reject H_0 if $-1.96 \leq t \leq 1.96$

$t(0.025, 213) \approx 1.96$ (z value)

Test Statistic: $t = \dfrac{(0.97774 - 1.0)}{0.01002} = -2.22$

4.7 ANOVA

Source	DF	SS	MS	F
Regression	3	300	100	10
Error (Residual)	27	270	10	
Total	30	570		

4.9 Before drawing conclusions from the small t statistics consider the F statistic for the overall fit of the model. Using a 5% level of significance, the test would be performed as follows:

Hypotheses: $H_0: \beta_1 = \beta_2 = \beta_3 = \beta_4 = \beta_5 = \beta_6 = \beta_7 = 0$
H_a: At least one of the coefficients is not equal to 0.

Decision Rule: Reject H_0 if $F > 2.17$ $F(0.05; 7, 82) \approx 2.17$
Do not reject H_0 if $F \leq 2.17$

Test Statistic: $F = 4.95$

OR

Decision Rule: Reject H_0 if p value < 0.05
Do not reject H_0 if p value ≥ 0.05

Test Statistic: p value = 0.000

Decision: Reject H_0

The conclusion from this test is that at least one of the coefficients is not equal to 0. In other words, at least one of the variables is helping to explain a significant amount of the variation in y. This conclusion contradicts the conclusion reached from examining each individual t statistic. When the overall F test is in conflict with the t tests, this suggests that multicollinearity is likely a problem.

4.11 Regression of FUELCON on DRIVERS, HWYMILES, GASTAX and INCOME

```
The regression equation is
FUELCON = 916 - 218 DRIVERS - 0.00078 HWYMILES - 3.69 GASTAX
          - 0.00549 INCOME


Predictor         Coef    SE Coef        T      P
Constant        916.10      73.27    12.50  0.000
DRIVERS        -218.18      62.13    -3.51  0.001
HWYMILES     -0.000776   0.001005    -0.77  0.444
GASTAX          -3.690      1.772    -2.08  0.043
INCOME       -0.005492   0.001767    -3.11  0.003


S = 56.2806    R-Sq = 44.4%    R-Sq(adj) = 39.6%


Analysis of Variance

Source             DF       SS      MS      F      P
Regression          4   116349   29087   9.18  0.000
Residual Error     46   145705    3168
Total              50   262054


Source      DF   Seq SS
DRIVERS      1    72943
HWYMILES     1      740
GASTAX       1    12063
INCOME       1    30602


Unusual Observations

Obs  DRIVERS  FUELCON      Fit   SE Fit   Residual   St Resid
 11     0.92   336.97   498.67    15.67    -161.70      -2.99R
 43     0.93   502.17   444.41    40.76      57.76       1.49 X
 50     0.67   715.55   553.97    17.59     161.58       3.02R
 51     1.38   289.99   318.49    34.08     -28.50      -0.64 X

R denotes an observation with a large standardized residual.
X denotes an observation whose X value gives it large influence.
```

a $\hat{y}_i = 916 - 218x_1 - 0.00078x_2 - 3.69x_3 - 0.00549x_4$

OR

FUELCON = 916 - 218 DRIVERS - 0.00078 HWYMILES

- 3.69 GASTAX - 0.00549 INCOME

b Hypotheses: $H_0: \beta_1 = \beta_2 = \beta_3 = \beta_4 = 0$
H_a: At least one of the coefficients ($\beta_1, \beta_2, \beta_3, \beta_4$) is not equal to 0.

Decision Rule: Reject H_0 if $F > 2.61$ $F(0.05; 4, 46) \approx 2.61$
Do not reject H_0 if $F \leq 2.61$

Test Statistic: $F = 9.18$

OR

Decision Rule: Reject H_0 if p value < 0.05
Do not reject H_0 if p value ≥ 0.05

Test Statistic: p value $= 0.000$

Decision: Reject H_0

Conclusion: At least one of the coefficients ($\beta_1, \beta_2, \beta_3, \beta_4$) is not equal to 0. In other words, at least one of the variables (x_1, x_2, x_3, x_4) is helping to explain a significant amount of the variation in y.

c 44.4%

d HWYMILES

Hypotheses: $H_0: \beta_2 = 0.0$
$H_a: \beta_2 \neq 0.0$

Decision Rule: Reject H_0 if $t > 1.96$ or $t < -1.96$
Do not reject H_0 if $-1.96 \leq t \leq 1.96$

$t(0.025, 46) \approx 1.96$ (z value)

Test Statistic: $t = -0.77$

OR

Decision Rule: Reject H_0 if p value < 0.05
Do not reject H_0 if p value ≥ 0.05

Test Statistic: p value $= 0.444$

Decision: Do not reject H_0

Conclusion: HWYMILES is not linearly related to FUELCON after taking account of the effect of the other variables. At this point, the regression should be re-run with HWYMILES omitted and the remaining variables should be reassessed.

e   Other possible variables might include the population in the state or the size of the state.

4.13 A time series plot of prime rate reveals no linear trend component that can be modeled. Because the series "wanders", using lagged values of the dependent variable as explanatory variables would be a logical choice.

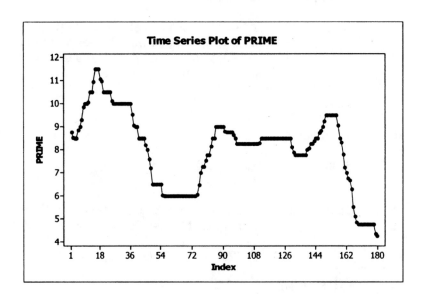

The one-period lagged prime rate produces a good fit (R-square = 98.4%)

The regression equation is
PRIME = - 0.0768 + 1.01 PRIMEL1

179 cases used, 1 case contains missing values

```
Predictor        Coef    SE Coef         T      P
Constant     -0.07683    0.08037     -0.96  0.340
PRIMEL1       1.00640    0.00976    103.11  0.000
```

S = 0.212187   R-Sq = 98.4%   R-Sq(adj) = 98.4%

Analysis of Variance

```
Source           DF       SS       MS         F      P
Regression        1   478.66   478.66  10631.49  0.000
Residual Error  177     7.97     0.05
Total           178   486.63
```

Unusual Observations

```
Obs  PRIMEL1    PRIME      Fit   SE Fit  Residual  St Resid
  8      9.3   9.8400   9.2727   0.0198    0.5673     2.69R
 12     10.1  10.5000  10.0375   0.0250    0.4625     2.19R
 14     10.5  10.9300  10.4904   0.0285    0.4396     2.09R
 15     10.9  11.5000  10.9232   0.0321    0.5768     2.75R
 18     11.5  11.0700  11.4968   0.0370   -0.4268    -2.04R
 20     11.0  10.5000  10.9735   0.0325   -0.4735    -2.26R
 37     10.0   9.5200   9.9872   0.0246   -0.4672    -2.22R
 38      9.5   9.0500   9.5041   0.0212   -0.4541    -2.15R
 41      9.0   8.5000   8.9808   0.0183   -0.4808    -2.27R
 49      7.2   6.5000   7.1793   0.0180   -0.6793    -3.21R
 55      6.5   6.0200   6.4648   0.0221   -0.4448    -2.11R
 76      6.1   6.4500   6.0220   0.0252    0.4280     2.03R
 77      6.5   6.9900   6.4145   0.0224    0.5755     2.73R
 83      7.8   8.1500   7.7228   0.0162    0.4272     2.02R
 86      8.5   9.0000   8.4776   0.0164    0.5224     2.47R
157      9.5   9.0500   9.4840   0.0211   -0.4340    -2.06R
158      9.1   8.5000   9.0311   0.0185   -0.5311    -2.51R
160      8.3   7.8000   8.2964   0.0160   -0.4964    -2.35R
161      7.8   7.2400   7.7731   0.0161   -0.5331    -2.52R
166      6.3   5.5300   6.2434   0.0236   -0.7134    -3.38R
180      4.4   4.2500   4.3010   0.0396   -0.0510    -0.24 X
```

R denotes an observation with a large standardized residual.
X denotes an observation whose X value gives it large influence.

The two-period lagged variable is still significant (p value = 0.000)

```
The regression equation is
PRIME = 0.0326 + 1.53 PRIMEL1 - 0.532 PRIMEL2


178 cases used, 2 cases contain missing values

Predictor        Coef   SE Coef        T       P     VIF
Constant      0.03258   0.06987     0.47   0.642
PRIMEL1       1.52680   0.06410    23.82   0.000    59.3
PRIMEL2      -0.53217   0.06504    -8.18   0.000    59.3

S = 0.180934     R-Sq = 98.8%    R-Sq(adj) = 98.8%

Analysis of Variance

Source             DF        SS       MS         F        P
Regression          2    480.69   240.34   7341.67    0.000
Residual Error    175      5.73     0.03
Total             177    486.42


Source     DF    Seq SS
PRIMEL1     1    478.50
PRIMEL2     1      2.19

Unusual Observations

Obs    PRIMEL1      PRIME       Fit    SE Fit   Residual   St Resid
  8        9.3     9.8400    9.4270    0.0252     0.4130       2.31R
 12       10.1    10.5000   10.0552    0.0214     0.4448       2.48R
 14       10.5    10.9300   10.4762    0.0244     0.4538       2.53R
 15       10.9    11.5000   11.1327    0.0373     0.3673       2.07R
 16       11.5    11.5000   11.7741    0.0461    -0.2741      -1.57 X
 18       11.5    11.0700   11.4708    0.0318    -0.4008      -2.25R
 19       11.1    10.9800   10.8143    0.0419     0.1657       0.94 X
 20       11.0    10.5000   10.9057    0.0291    -0.4057      -2.27R
 21       10.5    10.5000   10.2207    0.0412     0.2793       1.59 X
 25       10.5    10.1100   10.4762    0.0244    -0.3662      -2.04R
 37       10.0     9.5200    9.9788    0.0211    -0.4588      -2.55R
 41        9.0     8.5000    8.9842    0.0156    -0.4842      -2.69R
 49        7.2     6.5000    7.0069    0.0262    -0.5069      -2.83R
 50        6.5     6.5000    6.1198    0.0463     0.3802       2.17RX
 55        6.5     6.0200    6.4977    0.0192    -0.4777      -2.65R
 78        7.0     7.2500    7.2724    0.0416    -0.0224      -0.13 X
 83        7.8     8.1500    7.7409    0.0140     0.4091       2.27R
 86        8.5     9.0000    8.4869    0.0141     0.5131       2.84R
157        9.5     9.0500    9.4815    0.0181    -0.4315      -2.40R
160        8.3     7.8000    8.2121    0.0173    -0.4121      -2.29R
166        6.3     5.5300    6.0713    0.0292    -0.5413      -3.03R
167        5.5     5.1000    5.1337    0.0502    -0.0337      -0.19 X
179        4.8     4.3500    4.7571    0.0315    -0.4071      -2.28R

R denotes an observation with a large standardized residual.
X denotes an observation whose X value gives it large influence.
```

The three-period lagged variable is also significant.

```
The regression equation is
PRIME = 0.0604 + 1.44 PRIMEL1 - 0.265 PRIMEL2 - 0.180 PRIMEL3

177 cases used, 3 cases contain missing values

Predictor        Coef   SE Coef       T       P      VIF
Constant      0.06043   0.06999    0.86   0.389
PRIMEL1       1.43705   0.07478   19.22   0.000     82.5
PRIMEL2       -0.2655    0.1312   -2.02   0.044    246.5
PRIMEL3      -0.18023   0.07608   -2.37   0.019     81.0

S = 0.178823    R-Sq = 98.9%    R-Sq(adj) = 98.8%

Analysis of Variance

Source            DF       SS       MS         F        P
Regression         3   480.68   160.23   5010.52    0.000
Residual Error   173     5.53     0.03
Total            176   486.21

Source    DF   Seq SS
PRIMEL1    1   478.29
PRIMEL2    1     2.21
PRIMEL3    1     0.18

Unusual Observations

Obs   PRIMEL1     PRIME      Fit   SE Fit   Residual   St Resid
  8       9.3    9.8400   9.4280   0.0249     0.4120      2.33R
 12      10.1   10.5000  10.0456   0.0215     0.4544      2.56R
 14      10.5   10.9300  10.5506   0.0400     0.3794      2.18R
 15      10.9   11.5000  11.0874   0.0416     0.4126      2.37R
 17      11.5   11.5000  11.5635   0.0507    -0.0635     -0.37 X
 18      11.5   11.0700  11.4608   0.0317    -0.3908     -2.22R
 25      10.5   10.1100  10.4695   0.0243    -0.3595     -2.03R
 37      10.0    9.5200   9.9738   0.0209    -0.4538     -2.56R
 41       9.0    8.5000   8.9734   0.0160    -0.4734     -2.66R
 49       7.2    6.5000   6.9673   0.0305    -0.4673     -2.65R
 50       6.5    6.5000   6.1210   0.0459     0.3790      2.19R
 51       6.5    6.5000   6.3761   0.0546     0.1239      0.73 X
 55       6.5    6.0200   6.5041   0.0192    -0.4841     -2.72R
 76       6.1    6.4500   6.0947   0.0229     0.3553      2.00R
 77       6.5    6.9900   6.6392   0.0354     0.3508      2.00R
 83       7.8    8.1500   7.7865   0.0239     0.3635      2.05R
 86       8.5    9.0000   8.5499   0.0303     0.4501      2.55R
157       9.5    9.0500   9.4781   0.0179    -0.4281     -2.41R
166       6.3    5.5300   6.0978   0.0313    -0.5678     -3.22R
167       5.5    5.1000   5.1379   0.0498    -0.0379     -0.22 X
168       5.1    4.8400   4.7894   0.0505     0.0506      0.29 X
179       4.8    4.3500   4.7693   0.0315    -0.4193     -2.38R

R denotes an observation with a large standardized residual.
X denotes an observation whose X value gives it large influence.
```

The four-period lagged variable is no help.

Also, multicolllinearity is becoming a problem at this point. The previous model with the one- two- and three-period lagged variables will be used.

```
The regression equation is
PRIME = 0.0812 + 1.42 PRIMEL1 - 0.294 PRIMEL2 - 0.014 PRIMEL3 -
0.119 PRIMEL4


176 cases used, 4 cases contain missing values


Predictor        Coef    SE Coef       T       P     VIF
Constant      0.08123    0.07116    1.14   0.255
PRIMEL1       1.41548    0.07595   18.64   0.000    85.3
PRIMEL2       -0.2938     0.1332   -2.21   0.029   254.6
PRIMEL3       -0.0139     0.1342   -0.10   0.917   252.5
PRIMEL4      -0.11889    0.07770   -1.53   0.128    82.5


S = 0.178586     R-Sq = 98.9%    R-Sq(adj) = 98.9%


Analysis of Variance

Source            DF        SS       MS        F       P
Regression         4    480.54   120.14  3766.84   0.000
Residual Error   171      5.45     0.03
Total            175    486.00


Source    DF   Seq SS
PRIMEL1    1   478.08
PRIMEL2    1     2.21
PRIMEL3    1     0.18
PRIMEL4    1     0.07


Unusual Observations

Obs   PRIMEL1    PRIME       Fit   SE Fit   Residual   St Resid
  5       8.5   8.8400    8.4567   0.0229     0.3833      2.16R
  8       9.3   9.8400    9.4532   0.0300     0.3868      2.20R
 12      10.1  10.5000   10.0597   0.0236     0.4403      2.49R
 14      10.5  10.9300   10.5300   0.0423     0.4000      2.31R
 15      10.9  11.5000   11.1265   0.0493     0.3735      2.18R
 16      11.5  11.5000   11.7535   0.0526    -0.2535     -1.49 X
 18      11.5  11.0700   11.5210   0.0509    -0.4510     -2.63R
 25      10.5  10.1100   10.4643   0.0245    -0.3543     -2.00R
 37      10.0   9.5200    9.9699   0.0211    -0.4499     -2.54R
 41       9.0   8.5000    8.9185   0.0390    -0.4185     -2.40R
 49       7.2   6.5000    6.9735   0.0309    -0.4735     -2.69R
 50       6.5   6.5000    6.1069   0.0468     0.3931      2.28R
 51       6.5   6.5000    6.3706   0.0549     0.1294      0.76 X
 52       6.5   6.5000    6.4244   0.0554     0.0756      0.45 X
```

```
 55     6.5    6.0200   6.5089   0.0195   -0.4889   -2.75R
 77     6.5    6.9900   6.6338   0.0356    0.3562    2.04R
 86     8.5    9.0000   8.5806   0.0362    0.4194    2.40R
157     9.5    9.0500   9.4755   0.0180   -0.4255   -2.39R
166     6.3    5.5300   6.0870   0.0321   -0.5570   -3.17R
167     5.5    5.1000   5.1684   0.0536   -0.0684   -0.40 X
179     4.8    4.3500   4.7783   0.0320   -0.4283   -2.44R
```

R denotes an observation with a large standardized residual.
X denotes an observation whose X value gives it large influence.

These forecasts were computed by hand, not by MINITAB. There may be some rounding differences if MINITAB (or some other computer package) is used. Forecasts for each month in 2003 were computed using the model

PRIME = 0.0604 + 1.44 PRIMEL1 - 0.265 PRIMEL2 - 0.180 PRIMEL3

Date	Forecast	Actual
1/03	PRIME = 0.0604 + 1.44(4.25) - 0.265(4.35) - 0.180(4.75) = 4.17	4.25
2/03	PRIME = 0.0604 + 1.44(4.17) - 0.265(4.25) - 0.180(4.35) = 4.16	4.25
3/03	PRIME = 0.0604 + 1.44(4.16) - 0.265(4.17) - 0.180(4.25) = 4.18	4.25
4/03	PRIME = 0.0604 + 1.44(4.18) - 0.265(4.16) - 0.180(4.17) = 4.23	4.25
5/03	PRIME = 0.0604 + 1.44(4.23) - 0.265(4.18) - 0.180(4.16) = 4.30	4.25
6/03	PRIME = 0.0604 + 1.44(4.30) - 0.265(4.23) - 0.180(4.18) = 4.38	4.22
7/03	PRIME = 0.0604 + 1.44(4.38) - 0.265(4.30) - 0.180(4.23) = 4.47	4.00
8/03	PRIME = 0.0604 + 1.44(4.47) - 0.265(4.38) - 0.180(4.30) = 4.56	4.00
9/03	PRIME = 0.0604 + 1.44(4.56) - 0.265(4.47) - 0.180(4.38) = 4.65	4.00
10/03	PRIME = 0.0604 + 1.44(4.65) - 0.265(4.56) - 0.180(4.47) = 4.74	4.00
11/03	PRIME = 0.0604 + 1.44(4.74) - 0.265(4.65) - 0.180(4.56) = 4.83	4.00
12/03	PRIME = 0.0604 + 1.44(4.83) - 0.265(4.74) - 0.180(4.65) = 4.92	4.00

Possible measures of forecast accuracy to be used here include MSD, MAD and MAPE.

4.15 **Step 1**: The initial model with all five candidate explanatory variables is run. The weakest variable (DEPEND) is chosen and removed. A new regression is run and the process is continued. The initial regression follows:

```
The regression equation is
ABSENT = 3.46 - 0.0165 COMPLX + 0.074 PAY - 0.0405 SENIOR
         - 0.0253 AGE - 0.041 DEPEND

Predictor        Coef     SE Coef       T       P
Constant         3.457      1.362    2.54   0.013
COMPLX       -0.016461   0.006865   -2.40   0.019
PAY             0.0742     0.1610    0.46   0.647
SENIOR        -0.04051    0.04782   -0.85   0.400
AGE           -0.02526    0.03243   -0.78   0.439
DEPEND         -0.0413     0.1232   -0.34   0.739


S = 1.37848    R-Sq = 18.9%    R-Sq(adj) = 13.2%


Analysis of Variance

Source           DF        SS       MS      F      P
Regression        5    31.449    6.290   3.31  0.010
Residual Error   71   134.915    1.900
Total            76   166.364


Source  DF    Seq SS
COMPLX   1    21.771
PAY      1     0.133
SENIOR   1     8.121
AGE      1     1.210
DEPEND   1     0.213


Unusual Observations

Obs  COMPLX  ABSENT     Fit   SE Fit  Residual  St Resid
 18    13.0   7.000   2.908    0.276     4.092     3.03R
 42    50.0   6.000   2.157    0.230     3.843     2.83R
 46    43.0   1.000   0.809    0.713     0.191     0.16 X
 51     8.0   0.000   2.825    0.423    -2.825    -2.15R
 54    23.0   6.000   2.710    0.296     3.290     2.44R
 71    67.0   1.000   0.399    0.691     0.601     0.50 X

R denotes an observation with a large standardized residual.
X denotes an observation whose X value gives it large influence.
```

Step 2: The regression with DEPEND removed is run and examined. PAY is identified as the weakest remaining variable.

Step 3: The regression with DEPEND and PAY removed is run and examined. SENIOR is identified as the weakest remaining variable.

Step 4: The regression with DEPEND, PAY and SENIOR removed is run and examined. Both COMPLX and AGE are significant at a 5% level of significance. This is the regression chosen as the final model and is shown below.

```
The regression equation is
ABSENT = 4.33 - 0.0162 COMPLX - 0.0452 AGE

Predictor          Coef     SE Coef        T        P
Constant         4.3267      0.7344     5.89    0.000
COMPLX        -0.016194    0.005941    -2.73    0.008
AGE            -0.04520     0.02181    -2.07    0.042


S = 1.35895    R-Sq = 17.9%    R-Sq(adj) = 15.6%

Analysis of Variance

Source              DF         SS         MS       F        P
Regression           2     29.704     14.852    8.04    0.001
Residual Error      74    136.660      1.847
Total               76    166.364


Source    DF   Seq SS
COMPLX     1   21.771
AGE        1    7.933


Unusual Observations

Obs   COMPLX   ABSENT     Fit   SE Fit   Residual   St Resid
 18     13.0    7.000   2.941    0.264      4.059      3.04R
 42     50.0    6.000   2.161    0.191      3.839      2.85R
 51      8.0    0.000   2.886    0.257     -2.886     -2.16R
 54     23.0    6.000   2.779    0.241      3.221      2.41R

R denotes an observation with a large standardized residual.
```

The signs of the coefficients suggest that absenteeism will decrease with increases in either job complexity or age. Management might consider employees with least complex jobs or younger employees (some of which may fall into both these categories) when trying to impact absenteeism rates.

Note that only 17.9% of the variation in absenteeism has been explained using the two-variable final model. This leaves 82.1% of the variation unexplained. Although factors related to absenteeism have been determined, there is much variation yet to be explained. However, in problems dealing with factors such as absenteeism, job satisfaction, etc., this is not uncommon. An R-square of 17.9% may be the best we can achieve. (Actually, a somewhat higher R-square can be achieved later in the text using techniques not yet available.)

4.17 Regression Using All Three Explanatory Variables

```
The regression equation is
WINS = - 21.9 + 0.0976 HR + 606 BA - 16.9 ERA


Predictor       Coef   SE Coef       T       P
Constant      -21.88     28.93   -0.76   0.456
HR           0.09759   0.03572    2.73   0.011
BA             606.3     100.8    6.02   0.000
ERA          -16.897     1.758   -9.61   0.000


S = 5.00226    R-Sq = 89.7%    R-Sq(adj) = 88.5%


Analysis of Variance

Source           DF       SS       MS       F       P
Regression        3   5661.6   1887.2   75.42   0.000
Residual Error   26    650.6     25.0
Total            29   6312.2


Source  DF   Seq SS
HR       1   1126.6
BA       1   2223.9
ERA      1   2311.1


Unusual Observations

Obs   HR    WINS      Fit   SE Fit   Residual   St Resid
  5  192  74.000   64.861    2.042      9.139      2.00R

R denotes an observation with a large standardized residual.
```

All three variables are useful in explaining the variation in WINS (using a 5% level of significance). Using the absolute value of the *t* statistics as a measure of relative importance, ERA would appear to be the most important variable, with BA second and HR third. Practically speaking, this suggests that teams should concentrate on quality pitching first, hitting second and power third. When will the Texas Rangers learn this?

4.19 Pairwise correlations may not be sufficient to say whether or not multicollinearity will be a problem. There may be more complex relationships that cannot be detected by pairwise correlations. For example x_1 may be correlated with a linear combination of x_2 and x_3.

CHAPTER 5
Fitting Curves to Data

5.1 Model Summaries:

	R-square	Adjusted R-square	Standard Error
Linear Model	99.2	99.0	14.34
Second Order Model	100.0	99.9	3.54

The second-order model appears better than the first-order model. The p value on the second-order term, RDSQR, is 0.000, so the second-order term is significant at any reasonable level of significance. The adjusted R-square is higher and the standard error is lower. The R-square on the second-order model is stated as 100%. Note that not all of the variance in the y variable is explained, but the R-square rounds off to 100%.

5.3   a $(0.0202)(3) = 0.0606$

   b $(1.96)(1/0.94) = 2.085$

   $(1.96)(1/0.95) = 2.063$

   $2.085 - 2.063 = 0.022$

   A rise in production volume from 0.94 to 0.95 of capacity results in a decrease in cost of 0.022, assuming INDEX remains constant.

   c   The following regression was used to determine the point prediction of COST when PROD = 0.87 (PRODINV = 1.15) and INDEX = 120. The resulting prediction is 4.6047.

d   The 95% prediction interval is (4.2354, 4.9740)

Model Using Inverse of Production Level with Forecast

The regression equation is
COST = - 0.070 + 1.96 PRODINV + 0.0202 INDEX

```
Predictor       Coef     SE Coef       T       P
Constant     -0.0701      0.2688   -0.26   0.797
PRODINV       1.9622      0.1597   12.28   0.000
INDEX       0.020152    0.002077    9.70   0.000
```

S = 0.169445   R-Sq = 95.5%   R-Sq(adj) = 95.0%

Analysis of Variance

```
Source            DF        SS        MS        F       P
Regression         2   10.4539    5.2270   182.05   0.000
Residual Error    17    0.4881    0.0287
Total             19   10.9420
```

```
Source     DF   Seq SS
PRODINV     1   7.7505
INDEX       1   2.7034
```

Unusual Observations

```
Obs   PRODINV     COST     Fit    SE Fit   Residual   St Resid
  5      2.00   6.4200  6.4741    0.1148    -0.0541     -0.43 X
```

X denotes an observation whose X value gives it large influence.

Predicted Values for New Observations

```
New
Obs     Fit   SE Fit       95% CI              95% PI
  1  4.6047   0.0439   (4.5121, 4.6974)   (4.2354, 4.9740)
```

Values of Predictors for New Observations

```
New
Obs   PRODINV   INDEX
  1      1.15     120
```

5.5 From the scatterplots, it appears that log transformations of both the dependent and explanatory variables are necessary.

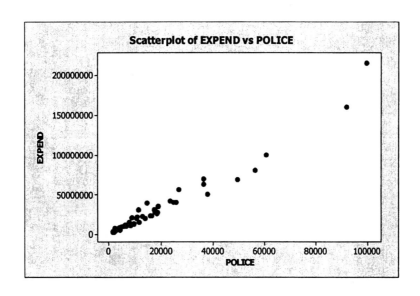

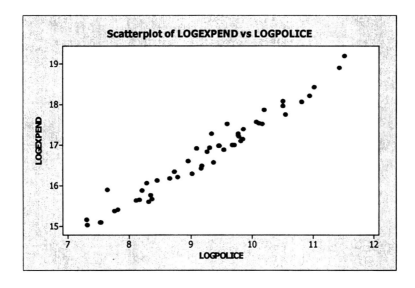

Regression Using Natural Logarithm of Dependent and Independent Variable

The regression equation is
LOGEXPEND = 8.19 + 0.927 LOGPOLICE

Predictor	Coef	SE Coef	T	P
Constant	8.1852	0.2451	33.40	0.000
LOGPOLICE	0.92725	0.02632	35.23	0.000

S = 0.199931   R-Sq = 96.2%   R-Sq(adj) = 96.1%

Analysis of Variance

Source	DF	SS	MS	F	P
Regression	1	49.609	49.609	1241.07	0.000
Residual Error	49	1.959	0.040		
Total	50	51.567			

Unusual Observations

Obs	LOGPOLICE	LOGEXPEND	Fit	SE Fit	Residual	St Resid
2	7.6	15.8923	15.2641	0.0509	0.6281	3.25R
24	9.3	17.2773	16.8410	0.0281	0.4363	2.20R
48	9.6	17.5087	17.0715	0.0293	0.4372	2.21R

R denotes an observation with a large standardized residual.

Regression with Prediction for x = Ln(10,000) = 9.21034

```
The regression equation is
LOGEXPEND = 8.19 + 0.927 LOGPOLICE

Predictor      Coef    SE Coef       T       P
Constant     8.1852     0.2451   33.40   0.000
LOGPOLICE   0.92725    0.02632   35.23   0.000

S = 0.199931     R-Sq = 96.2%    R-Sq(adj) = 96.1%

Analysis of Variance

Source            DF       SS        MS         F       P
Regression         1   49.609    49.609   1241.07   0.000
Residual Error    49    1.959     0.040
Total             50   51.567

Unusual Observations

Obs   LOGPOLICE   LOGEXPEND       Fit   SE Fit   Residual   St Resid
  2         7.6     15.8923   15.2641   0.0509     0.6281       3.25R
 24         9.3     17.2773   16.8410   0.0281     0.4363       2.20R
 48         9.6     17.5087   17.0715   0.0293     0.4372       2.21R

R denotes an observation with a large standardized residual.

Predicted Values for New Observations

New
Obs       Fit    SE Fit         95% CI              95% PI
  1   16.7255    0.0280   (16.6692, 16.7818)   (16.3198, 17.1312)

Values of Predictors for New Observations

New
Obs   LOGPOLICE
  1        9.21
```

To transform back to the original units: $\hat{y} = e^{16.7255}$. The predicted expenditure would be $18,356,604.76.

5.7

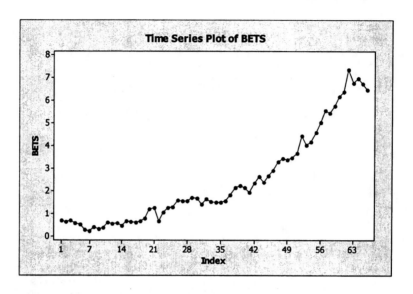

Regression of BETS on Linear Trend Variable (TIME)

The regression equation is
BETS = - 0.877 + 0.0985 TIME

```
Predictor        Coef     SE Coef       T      P
Constant       -0.8768     0.1977    -4.43  0.000
TIME          0.098493   0.005131    19.20  0.000


S = 0.794057   R-Sq = 85.2%   R-Sq(adj) = 85.0%

Analysis of Variance

Source          DF      SS       MS       F      P
Regression       1   232.36   232.36  368.52  0.000
Residual Error  64    40.35     0.63
Total           65   272.71

Unusual Observations

Obs  TIME   BETS      Fit   SE Fit  Residual  St Resid
 62  62.0  7.3500  5.2298   0.1759    2.1202     2.74R
```

R denotes an observation with a large standardized residual.

Regression of BETS on Linear and Second-Order Trend Variable

```
The regression equation is
BETS = 0.795 - 0.0490 TIME + 0.00220 TIMESQR


Predictor        Coef     SE Coef       T       P
Constant       0.7947      0.1240    6.41   0.000
TIME        -0.048991    0.008539   -5.74   0.000
TIMESQR      0.0022012   0.0001235   17.82   0.000


S = 0.325633    R-Sq = 97.6%    R-Sq(adj) = 97.5%

Analysis of Variance

Source            DF      SS       MS         F       P
Regression         2   266.03   133.02   1254.44   0.000
Residual Error    63     6.68     0.11
Total             65   272.71


Source    DF   Seq SS
TIME       1   232.36
TIMESQR    1    33.67


Unusual Observations

Obs   TIME    BETS     Fit   SE Fit   Residual   St Resid
 62   62.0  7.3500  6.2189   0.0910     1.1311      3.62R
 66   66.0  6.4700  7.1499   0.1167    -0.6799     -2.24R

R denotes an observation with a large standardized residual.
```

Using the quadratic trend model provides a better fit. The following measures of fit support the use of the quadratic trend.

	R-square	Adjusted R-square	Standard Error
Linear Model	85.2%	85.0%	0.794057
Second-Order Model	97.6%	97.5%	0.325633

Forecast for 1993:   BETS = 0.795 - 0.0490(67) + 0.00220(4489) = 7.3878

Forecast for 1994:   BETS = 0.795 - 0.0490(68) + 0.00220(4624) = 7.6358

CHAPTER 6
Assessing the Assumptions of the Regression Model

6.1 **a** Yes. It is not, however, a linear relationship.

b $\hat{y} = 10$

c Decision Rule: Reject H_0 if $t > 3.25$ or $t < -3.25$ $t(0.005, 9) = 3.25$
Do not reject H_0 if $-3.25 \leq t \leq 3.25$

Test Statistic: $t = 0.0$

OR

Decision Rule: Reject H_0 if p value < 0.01
Do not reject H_0 if p value ≥ 0.01

Test Statistic: p value $= 1.0$

Decision: Do not reject H_0

Conclusion: There is no <u>linear</u> relationship between y and x.

d There is a "strong" association. In fact, this is an exact relationship, but not a linear one. The equation expressing the relationship is: $y = x^2$

6.3 There is evidence that the constant variance assumption has been violated. The cone-shaped pattern that is typical of instances when this assumption is violated is evident in the residual plot in Figure 6.30. One possible correction would be to use the log of the monthly prices as the dependent variable. The regression results and the residual plots from this regression are shown. The correction appears to have eliminated the problem of nonconstant variance.

<u>Regression Using Log of Monthly Prices as Dependent Variable</u>

```
The regression equation is
LOGS&P = 0.0186 + 0.999 LOGS&P LAG1

346 cases used, 1 case contains missing values

Predictor        Coef     SE Coef        T        P
Constant       0.01856     0.01358     1.37    0.173
LOGS&P LAG1   0.998552    0.002082   479.52    0.000

S = 0.0465387     R-Sq = 99.9%     R-Sq(adj) = 99.9%

Analysis of Variance

Source            DF       SS        MS          F        P
Regression         1    498.02    498.02   229942.00    0.000
Residual Error   344      0.75      0.00
Total            345    498.77

Unusual Observations

        LOGS&P
Obs      LAG1    LOGS&P      Fit     SE Fit   Residual   St Resid
  7      4.53    4.43509   4.53743  0.00466   -0.10234     -2.21R
  8      4.44    4.31273   4.44723  0.00482   -0.13450     -2.91R
  9      4.31    4.46810   4.32505  0.00504    0.14306      3.09R
 12      4.40    4.52204   4.41449  0.00487    0.10755      2.32R
 24      4.72    4.83367   4.73051  0.00432    0.10316      2.23R
 57      4.97    4.88200   4.98458  0.00390   -0.10258     -2.21R
 74      5.15    5.05130   5.16467  0.00362   -0.11338     -2.44R
103      5.22    5.33943   5.23585  0.00352    0.10358      2.23R
105      5.35    5.46076   5.36263  0.00334    0.09813      2.11R
127      5.66    5.76485   5.67045  0.00295    0.09440      2.03R
152      6.26    6.17614   6.27192  0.00252   -0.09578     -2.06R
156      6.23    6.35670   6.23991  0.00253    0.11679      2.51R
165      6.54    6.29423   6.54586  0.00252   -0.25163     -5.41R
166      6.29    6.20823   6.30368  0.00251   -0.09545     -2.05R
199      6.74    6.64296   6.74651  0.00259   -0.10354     -2.23R
215      6.84    6.94486   6.84523  0.00265    0.09962      2.14R
295      8.09    7.93724   8.10024  0.00430   -0.16300     -3.52R
325      8.32    8.22717   8.32929  0.00470   -0.10212     -2.21R
344      7.95    7.83062   7.95273  0.00406   -0.12211     -2.63R

R denotes an observation with a large standardized residual.
```

Residual Plots from Transformed Regression

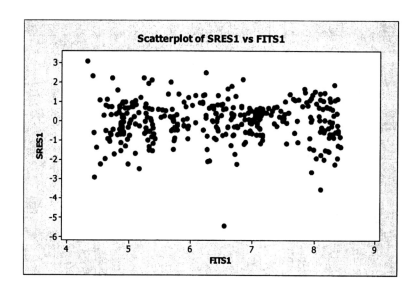

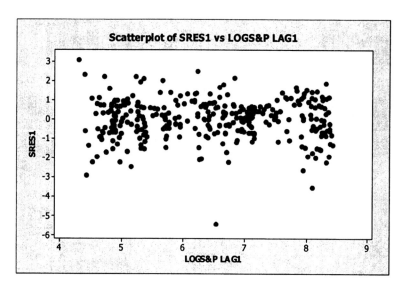

6.5 Hypotheses: $H_0: \rho = 0.0$
$H_a: \rho > 0.0$

Decision Rule: Reject H_0 if $d < d_L$ (0.05;27,2) = 1.24
Do not reject H_0 if $d > d_U$ (0.05; 27,2) = 1.56
Inconclusive if $1.24 \leq d \leq 1.56$

Test Statistic: $d = 2.14$

Decision: Do not reject H_0

Conclusion: The disturbances are NOT autocorrelated.

6.7 Regression of IMPORTS on GDP

```
The regression equation is
IMPORTS = 22.3 + 0.106 GDP

Predictor        Coef     SE Coef         T        P
Constant        22.32       19.24      1.16    0.258
GDP          0.105671    0.008452     12.50    0.000

S = 87.0031    R-Sq = 87.2%    R-Sq(adj) = 86.6%

Analysis of Variance

Source              DF          SS         MS         F        P
Regression           1     1183304    1183304    156.32    0.000
Residual Error      23      174099       7570
Total               24     1357403

Unusual Observations

Obs     GDP   IMPORTS       Fit    SE Fit   Residual   St Resid
 13    2660      53.8     303.4      22.5     -249.6     -2.97R
 25   10082    1148.0    1087.7      79.0       60.3      1.65 X

R denotes an observation with a large standardized residual.
X denotes an observation whose X value gives it large influence.
```

Residual Plots

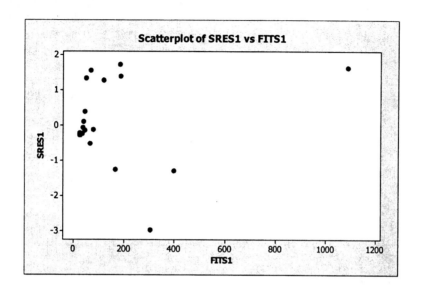

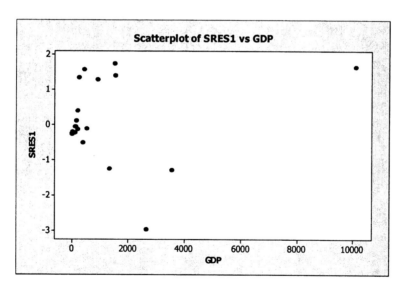

a   The residuals do not appear to be randomly distributed, indicating that some assumption has been violated. It may not be clear from the residual plots which assumption has been violated, however. It is actually the linearity assumption that is violated. In this example, the violation of the linearity assumption may be mistaken as a problem with outliers. There are obviously one or two extreme points in the data set. One is the United States, which has much greater IMPORTS and GDP than the other countries in the data set. If the violation of the linearity assumption was mistaken as an outlier problem, one possible course of action would be to remove the outlier (the U.S.) and see if the model could be improved. This is done in b).

b

Scatterplot with U.S. Removed from Data Set

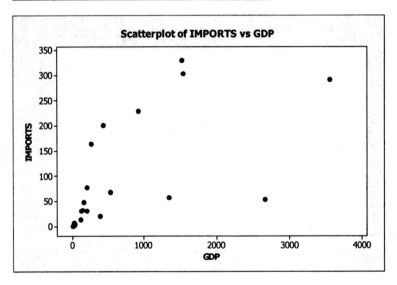

Regression with U.S. Removed from Data Set

```
The regression equation is
IMPORTS = 37.1 + 0.0763 GDP

Predictor       Coef    SE Coef       T      P
Constant       37.15      20.38    1.82  0.082
GDP          0.07629    0.01892    4.03  0.001

S = 83.5283   R-Sq = 42.5%   R-Sq(adj) = 39.9%

Analysis of Variance

Source             DF       SS      MS      F      P
Regression          1   113387  113387  16.25  0.001
Residual Error     22   153494    6977
Total              23   266881

Unusual Observations

Obs   GDP   IMPORTS    Fit   SE Fit  Residual  St Resid
 13  2660      53.8  240.1     42.7    -186.3   -2.60RX
 16  3550     292.1  308.0     58.5     -15.9   -0.27 X
 24  1520     330.1  153.1     24.5     177.0    2.22R

R denotes an observation with a large standardized residual.
X denotes an observation whose X value gives it large influence.
```

Residual Plots for Regression with U.S. Removed from Data Set

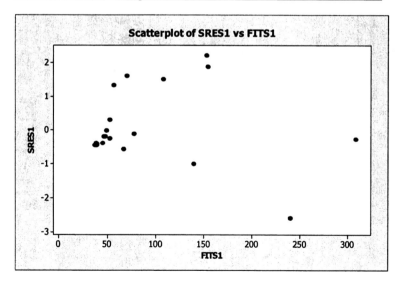

Only the plot of the standardized residuals versus the fitted values is included here. The pattern in the plot of the standardized residuals versus GDP is nearly identical to the pattern in this plot.

c   The removal of the U.S. from the data set obviously does not help improve the regression. This can be seen from the residual plots from the regression excluding the U.S. The problem in this data set is the large differences between the IMPORTS figures and between the GDP figures for the different countries, leading to a curvilinear pattern in the data. A possible correction that would narrow these large differences would be to use the natural logarithm transformation on both the dependent and explanatory variable.

Scatterplot Using Logs of Imports and GDP (U.S. Included)

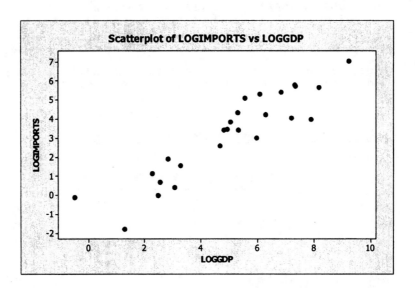

Regression Using Logs of Imports and GDP

```
The regression equation is
LOGIMPORTS = - 1.13 + 0.867 LOGGDP


Predictor        Coef   SE Coef       T       P
Constant      -1.1275    0.4346   -2.59   0.016
LOGGDP        0.86703   0.07877   11.01   0.000


S = 0.914202    R-Sq = 84.0%    R-Sq(adj) = 83.4%

Analysis of Variance

Source            DF       SS       MS        F       P
Regression         1   101.26   101.26   121.15   0.000
Residual Error    23    19.22     0.84
Total             24   120.48

Unusual Observations

Obs   LOGGDP   LOGIMPORTS      Fit   SE Fit   Residual   St Resid
 17     1.28       -1.772   -0.017    0.346     -1.755     -2.07R
 23    -0.48       -0.105   -1.545    0.469      1.439      1.83 X

R denotes an observation with a large standardized residual.
X denotes an observation whose X value gives it large influence.
```

Residual Plot from the Regression Using Logs of Imports and GDP

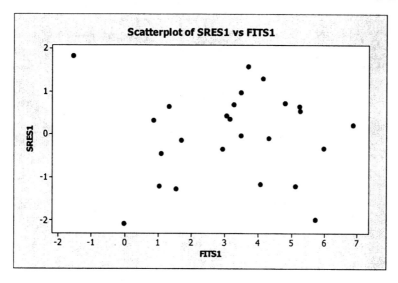

Only the plot of the standardized residuals versus the fitted values is included here. The pattern in the plot of the standardized residuals versus the LOGGDP variable is nearly identical to the pattern in this plot. From the residual plots of the regression of LOGIMP on LOGGDP, this appears to be the preferred regression. The residuals appear to be randomly scattered around their mean in the plots. Recall that this regression cannot be compared to the original using R-square or s_e, however, because the dependent variable has been transformed.

6.9 No. You cannot compare these two models on the basis of the traditional summary statistics like R-square or standard error because the dependent variable has been transformed. To choose the preferred model, residual plots could be used. The model which appears to most closely conform to the assumptions according to residual plots would be the one chosen.

6.11 From the residual plots, it appears that the constant variance assumption has been violated. The residuals exhibit the cone-shaped pattern that indicates an increase in the variance as the explanatory variable (CUTTING) increases. Two possible corrections for nonconstant variance are:

1) $\ln(\text{FATALS}) = \beta_0 + \beta_1 \text{CUTTING} + e$

2) $\text{FATALS/CUTTING} = \beta_0 (1/\text{CUTTING}) + \beta_1 + e$

The first of these two corrections will be illustrated here.

Note that four of the values of FATALS are outside the range of the natural logarithm transformation (and are treated as missing data). In this case the values outside the range are 0s. There are various ways of handling this problem so that the natural logarithm transformation can still be used. In this problem, the 0 values were basically ignored. MINITAB will automatically assign a missing value code when a value is outside the range of a transformation. The resulting missing cases will not be used in any subsequent analyses. That is what was done in the solution that follows. This is not the most eloquent way to approach this problem, but will probably supply reasonable answers in this situation.

One other possible solution is to increment each value of the variable FATAL by some small amount (0.01, for example), so that no 0 values are present. When this is tried in this example, some outliers are produced, which further complicates the analysis. I do not believe the regression results obtained from the incremented data are reliable.

One alternative I did try in this situation was to change the 0s to 1s (reasoning that 1 and 0 fatalities were fairly similar numbers). The results in this case were very similar to the ones shown here, with the 0s omitted, and might be a better way to approach the problem. The outlier problem is not present when this larger increment is used.

Regression of FATALS on CUTTING

```
The regression equation is
FATALS = - 47.7 + 0.0134 CUTTING

Predictor        Coef     SE Coef       T       P
Constant       -47.71       23.17   -2.06   0.044
CUTTING      0.013432    0.001903    7.06   0.000

S = 86.6240     R-Sq = 44.6%    R-Sq(adj) = 43.7%

Analysis of Variance

Source            DF        SS       MS       F       P
Regression         1    373953   373953   49.84   0.000
Residual Error    62    465230     7504
Total             63    839184


Unusual Observations

Obs   CUTTING   FATALS     Fit   SE Fit   Residual   St Resid
 10     18660    486.0   202.9     18.5      283.1      3.35R
 12     17466    373.0   186.9     16.7      186.1      2.19R
 14     15261    347.0   157.3     13.8      189.7      2.22R
 26     12342    292.0   118.1     11.2      173.9      2.03R

R denotes an observation with a large standardized residual.
```

Residual Plots from Linear Regression

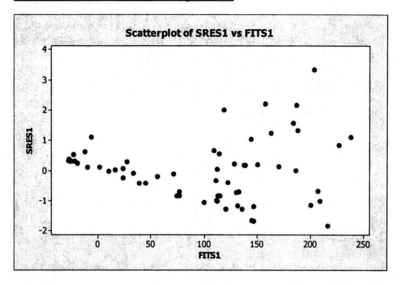

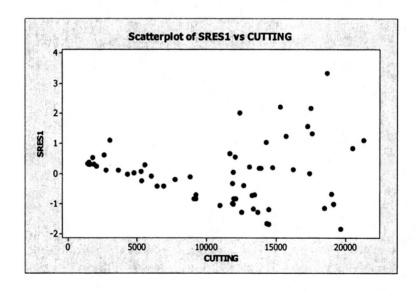

Regression of LN(FATALS) on CUTTING

The regression equation is
LOGFATAL = 1.55 + 0.000195 CUTTING

60 cases used, 4 cases contain missing values

```
Predictor          Coef      SE Coef       T       P
Constant         1.5493       0.3564    4.35   0.000
CUTTING      0.00019493   0.00002835    6.87   0.000
```

S = 1.17952   R-Sq = 44.9%   R-Sq(adj) = 44.0%

Analysis of Variance

```
Source           DF        SS        MS        F       P
Regression        1    65.758    65.758    47.26   0.000
Residual Error   58    80.693     1.391
Total            59   146.451
```

Unusual Observations

```
Obs   CUTTING   LOGFATAL    Fit   SE Fit   Residual   St Resid
 35    14424      1.099    4.361   0.175    -3.262     -2.80R
 36    14315      1.099    4.340   0.174    -3.241     -2.78R
 41     9054      0.693    3.314   0.166    -2.621     -2.24R
 54     3060      4.477    2.146   0.280     2.332      2.04R
```

R denotes an observation with a large standardized residual.

Residual Plots from Transformed Regression

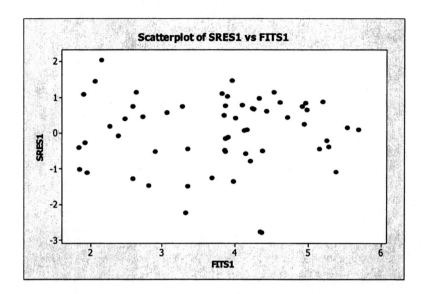

Only the plot of the standardized residuals versus the fitted values is included here. The pattern in the plot of the standardized residuals versus the CUTTING variable is nearly identical to the pattern in this plot.

6.13 The scatterplot of the dependent variable (TIME) versus each of the explanatory variables (NUMBER and EXPER), the regression and the associated residual plots follow. This is the output necessary to determine whether any assumptions have been violated. From the residual plots it is clear that an assumption has been violated: the linearity assumption. The plot of the standardized residuals (SRES1) versus NUMBER indicates that the explanatory variable NUMBER is the one that should enter the equation in a curvilinear manner. The residual plot of SRES1 versus EXPER does not indicate any problems. Looking back at the scatterplot of TIME versus NUMBER suggests that an appropriate correction might be to use a second-order polynomial regression with both NUMBER and NUMBER^2 (with the second-order variable called NUMSQR). The second-order regression of TIME on NUMBER, NUMSQR, and EXPER is shown.

The residual plots from the second-order regression suggest that the new model is an improvement over the original. The residuals in these plots appear to be randomly distributed. The R-square and adjusted R-square have both increased, and the standard error of the regression has decreased. These are all positive signs in favor of the new model. A *t* test for the coefficient of the variable EXPER, however, will show that this variable is not important in the regression. Thus, it has been removed and a final regression has been run using the explanatory variables NUMBER and NUMSQR. This is the preferred model.

Scatterplot of TIME versus Each Explanatory Variable

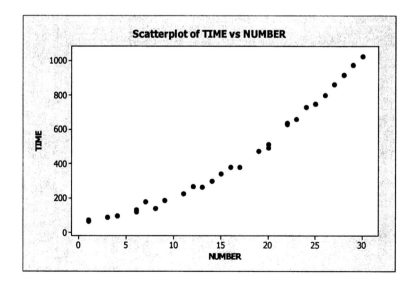

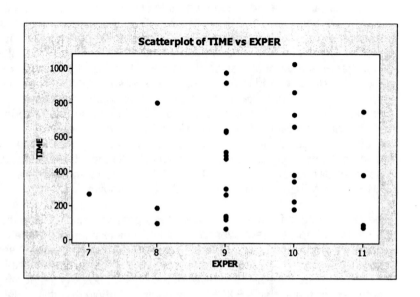

Regression of TIME on NUMBER and EXPER

```
The regression equation is
TIME = - 179 + 33.0 NUMBER + 10.2 EXPER


Predictor     Coef   SE Coef       T      P
Constant    -179.3     125.6   -1.43  0.165
NUMBER      32.969     1.436   22.96  0.000
EXPER        10.19     13.12    0.78  0.444


S = 68.4118    R-Sq = 95.1%    R-Sq(adj) = 94.8%

Analysis of Variance

Source           DF        SS       MS       F      P
Regression        2   2474190  1237095  264.33  0.000
Residual Error   27    126365     4680
Total            29   2600555


Source     DF    Seq SS
NUMBER      1   2471367
EXPER       1      2823
```

Residual Plots from Regression of TIME on NUMBER and EXPER

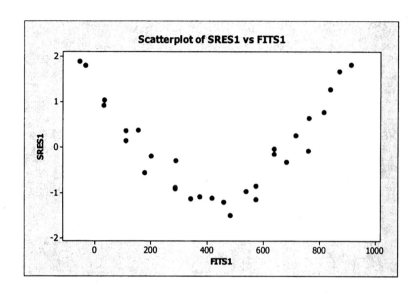

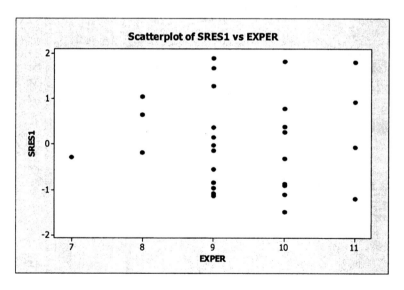

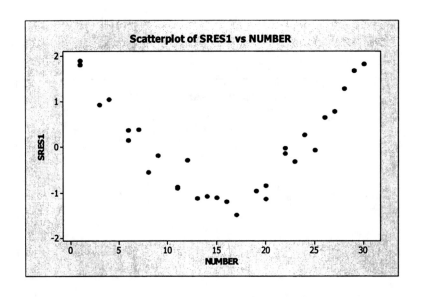

Second-Order Regression of TIME on NUMBER, NUMSQR and EXPER

```
The regression equation is
TIME = 65.7 + 3.89 NUMBER + 0.943 NUMSQR + 0.37 EXPER

Predictor        Coef    SE Coef       T      P
Constant        65.73      29.77    2.21  0.036
NUMBER          3.887      1.308    2.97  0.006
NUMSQR        0.94317    0.04114   22.93  0.000
EXPER           0.371      2.934    0.13  0.900

S = 15.1350    R-Sq = 99.8%    R-Sq(adj) = 99.7%

Analysis of Variance

Source           DF        SS       MS        F      P
Regression        3   2594599   864866  3775.58  0.000
Residual Error   26      5956      229
Total            29   2600555

Source    DF    Seq SS
NUMBER     1   2471367
NUMSQR     1    123228
EXPER      1         4

Unusual Observations

Obs  NUMBER    TIME     Fit   SE Fit  Residual  St Resid
  7     7.0  178.00  142.87     4.31     35.13      2.42R

R denotes an observation with a large standardized residual.
```

Residual Plots from Second-Order Regression

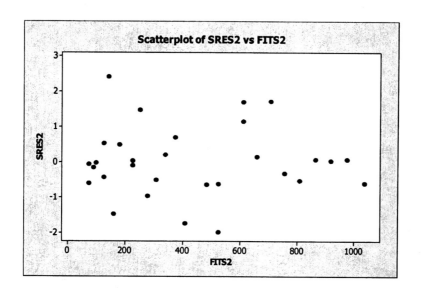

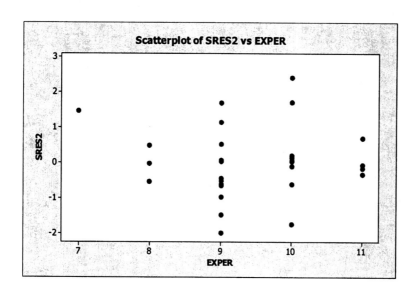

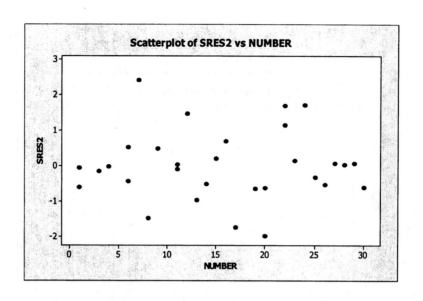

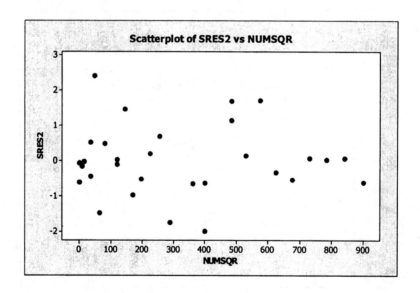

Final Regression Omitting EXPER

```
The regression equation is
TIME = 69.3 + 3.86 NUMBER + 0.944 NUMSQR


Predictor       Coef    SE Coef        T       P
Constant      69.330      8.551     8.11   0.000
NUMBER         3.865      1.271     3.04   0.005
NUMSQR       0.94393    0.03995    23.63   0.000


S = 14.8567    R-Sq = 99.8%    R-Sq(adj) = 99.8%

Analysis of Variance

Source            DF         SS        MS        F       P
Regression         2    2594595   1297298  5877.56   0.000
Residual Error    27       5959       221
Total             29    2600555


Source   DF    Seq SS
NUMBER    1   2471367
NUMSQR    1    123228


Unusual Observations

Obs  NUMBER    TIME     Fit   SE Fit  Residual  St Resid
  7     7.0  178.00  142.64     3.83     35.36      2.46R
 20    20.0  495.00  524.20     3.74    -29.20     -2.03R

R denotes an observation with a large standardized residual.
```

Note: To save space, the residual plots from the final model have not been included. However, to complete the analysis the following plots should be done: standardized residuals versus fitted values, standardized residuals versus each explanatory variable in the model, and standardized residuals versus each possible explanatory variable excluded from the model. Examining these plots reveals no further assumptions violated.

6.15 <u>Regression of WINS on HR, BA and ERA</u>

```
The regression equation is
WINS = - 21.9 + 0.0976 HR + 606 BA - 16.9 ERA

Predictor        Coef    SE Coef       T      P
Constant       -21.88      28.93   -0.76  0.456
HR            0.09759    0.03572    2.73  0.011
BA              606.3      100.8    6.02  0.000
ERA           -16.897      1.758   -9.61  0.000

S = 5.00226   R-Sq = 89.7%   R-Sq(adj) = 88.5%

Analysis of Variance

Source          DF       SS      MS      F      P
Regression       3   5661.6  1887.2  75.42  0.000
Residual Error  26    650.6    25.0
Total           29   6312.2

Source   DF   Seq SS
HR        1   1126.6
BA        1   2223.9
ERA       1   2311.1

Unusual Observations

Obs   HR    WINS      Fit   SE Fit  Residual  St Resid
  5  192  74.000   64.861    2.042     9.139      2.00R

R denotes an observation with a large standardized residual.
```

Residual Plots for Linear Regression

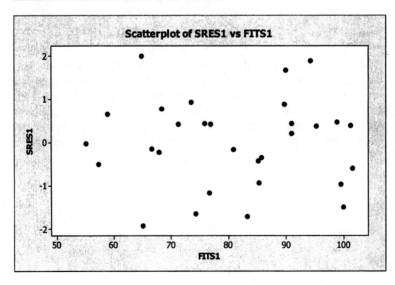

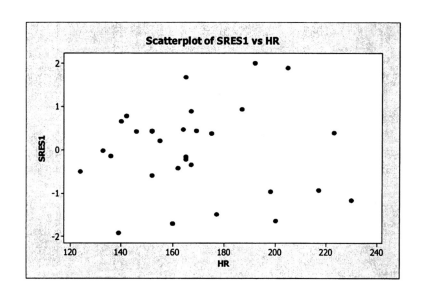

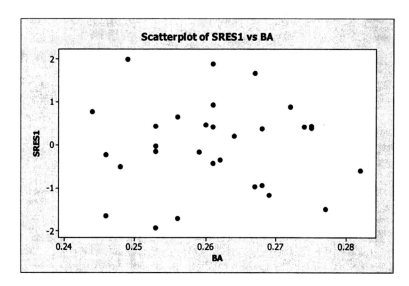

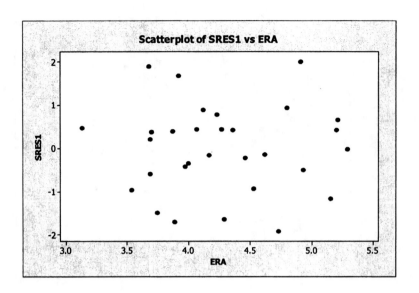

There do not appear to be any assumptions violated.

6.17 The scatterplot of the dependent variable (VALUE) versus each of the explanatory variables (SIZE, DEPRECIATION, and CONDITION), the regression, and the associated residual plots follow. This is the output necessary to determine whether any assumptions have been violated. From the residual plots, it is clear that an assumption has been violated. In this case it is the linearity assumption. To correct for the violation, logs of the dependent variable, VALUE, and two of the independent variables, DEPRECIATION and SIZE, are used. The revised regression output, along with residual plots, is shown. In the revised regression, the variable CONDITION is not significant. The final model is obtained by deleting this variable.

The residual plots from the revised regression suggest that the new model is an improvement over the original. The residuals in these plots appear to be randomly distributed. The R-square and adjusted R-square cannot be compared, because the dependent variable has been transformed. However, from the residual plots we can conclude that the transformed model is preferred to the original.

It is easy in this problem to mistake the nonlinearity for a violation of the constant variance assumption. (I conclude that it is easy because I did it.) However, after trying a new regression with the log of VALUE as the dependent variable to correct for the nonconstant variance, it will become obvious that something is still wrong. The correct transformation is to take logs of both the dependent and independent variables.

The independent variable CONDITION was not transformed using logs. First, this variable has a lot of 0 values, so we would be faced with the problem of some additional type of change because the log of 0 is not defined. Also, even if we make such a change, it seems to be of little help. The variable CONDITION simply does not seem to be useful in this model.

An equation similar to the one developed here could be used to value homes based on the known characteristics of the houses. For a 1400 square foot house with physical condition index 0.02 and depreciation factor 0.7, the equation would produce a valuation of:

Logvalue = 4.25 + 1.01(7.244) + 1.57(-0.36) = 11

Note that natural logarithms of the independent variables have been used in the equation and the resulting value of the dependent variable is the natural log of the value of the house. This should be converted back into dollar units:

$\hat{y} = e^{11} = \$59{,}874.14$

Also note that the depreciation factor was not used in the equation because it was determined that this variable was not related to value. An equation of this type would allow a quick estimate of the value of the house and would avoid the need for regular in-person assessments. Obviously, other variables could be included as well as size and depreciation factor, if available and judged to be important.

Scatterplot of Value versus Each Explanatory Variable

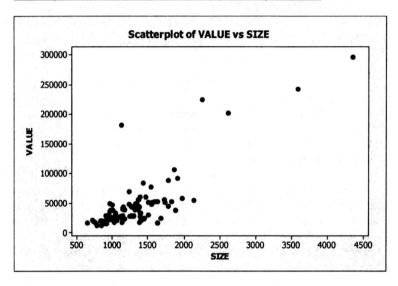

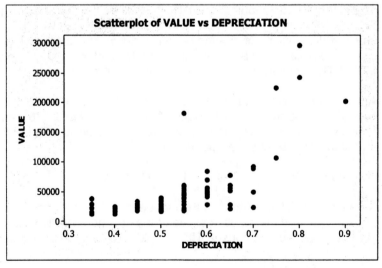

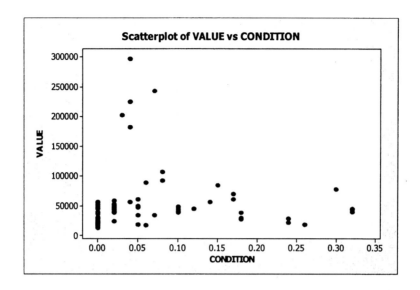

Regression of VALUE on All Three Explanatory Variables

```
The regression equation is
VALUE = - 105944 + 55.8 SIZE + 146276 DEPRECIATION - 4557
CONDITION

Predictor        Coef    SE Coef       T      P
Constant      -105944      13051   -8.12  0.000
SIZE           55.811      5.841    9.55  0.000
DEPRECIATION   146276      30118    4.86  0.000
CONDITION       -4557      31112   -0.15  0.884


S = 24509.8    R-Sq = 73.5%    R-Sq(adj) = 72.6%


Analysis of Variance

Source          DF            SS            MS       F      P
Regression       3   1.59668E+11   53222738475   88.60  0.000
Residual Error  96   57670074110     600729939
Total           99   2.17338E+11


Source          DF        Seq SS
SIZE             1   1.44459E+11
DEPRECIATION     1   15196378226
CONDITION        1      12888976


Unusual Observations

Obs  SIZE   VALUE     Fit   SE Fit  Residual  St Resid
 52  1148   43566   37121     8511      6445      0.28 X
 53  1363   38950   49120     8609    -10170     -0.44 X
```

```
54   1262    44633    43483     8540     1150     0.05 X
76   2251   224182   129212     6007    94970     4.00R
77   1126   182012    37169     2888   144843     5.95R
78   2617   201597   171625     9357    29972     1.32 X
97   3581   242690   210617    10894    32073     1.46 X
98   4343   296251   253282    14628    42969     2.18RX
```

R denotes an observation with a large standardized residual.
X denotes an observation whose X value gives it large influence.

Residual Plots from Original Regression

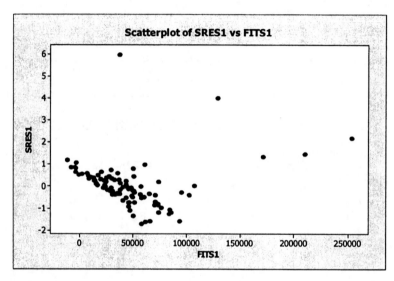

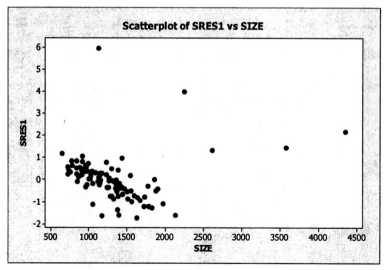

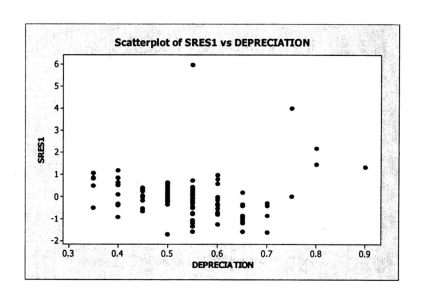

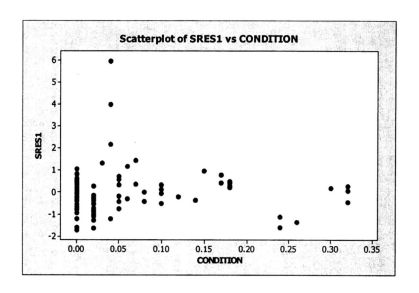

Regression Using Log Transformed Variable (and CONDITION)

```
The regression equation is
LOGVALUE = 4.14 + 1.02 LOGSIZE + 1.52 LOGDEP + 0.325 CONDITION


Predictor      Coef    SE Coef      T       P
Constant     4.1425     0.9355    4.43   0.000
LOGSIZE      1.0237     0.1201    8.52   0.000
LOGDEP       1.5220     0.2062    7.38   0.000
CONDITION    0.3251     0.4150    0.78   0.435


S = 0.327302    R-Sq = 74.5%    R-Sq(adj) = 73.7%


Analysis of Variance

Source           DF       SS       MS       F       P
Regression        3    30.077   10.026   93.59   0.000
Residual Error   96    10.284    0.107
Total            99    40.361


Source       DF   Seq SS
LOGSIZE       1   23.293
LOGDEP        1    6.718
CONDITION     1    0.066


Unusual Observations

Obs   LOGSIZE   LOGVALUE     Fit   SE Fit   Residual   St Resid
  3      7.40     9.7280  10.6602  0.0530    -0.9322     -2.89R
 36      7.54    10.5361  10.2654  0.1198     0.2707      0.89 X
 46      7.24    10.2454  10.9747  0.0829    -0.7293     -2.30R
 47      6.96     9.9496  10.6939  0.0899    -0.7443     -2.37R
 48      7.23     9.7898  10.7144  0.0914    -0.9246     -2.94R
 52      7.05    10.6820  10.5492  0.1136     0.1328      0.43 X
 53      7.22    10.5700  10.7249  0.1148    -0.1549     -0.51 X
 54      7.14    10.7062  10.6461  0.1138     0.0601      0.20 X
 62      7.06    10.0623  10.8382  0.0752    -0.7758     -2.44R
 76      7.72    12.3202  11.6195  0.0784     0.7007      2.20R
 77      7.03    12.1118  10.4384  0.0384     1.6735      5.15R
 97      8.18    12.3995  12.2028  0.1139     0.1968      0.64 X
 98      8.38    12.5990  12.3905  0.1324     0.2084      0.70 X

R denotes an observation with a large standardized residual.
X denotes an observation whose X value gives it large influence.
```

Residual Plots from Transformed Model

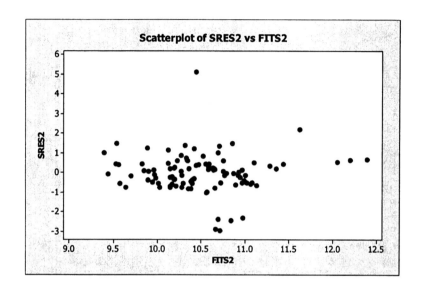

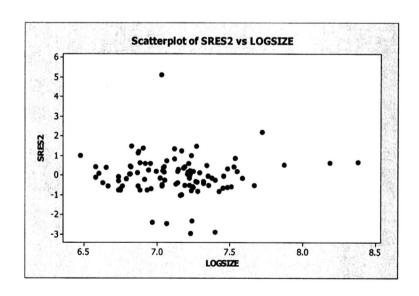

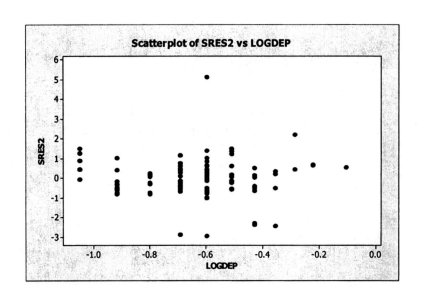

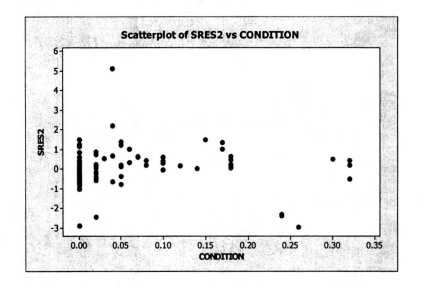

Regression after Deleting CONDITION

```
The regression equation is
LOGVALUE = 4.25 + 1.01 LOGSIZE + 1.57 LOGDEP

Predictor     Coef    SE Coef      T       P
Constant    4.2532     0.9229    4.61   0.000
LOGSIZE     1.0146     0.1193    8.51   0.000
LOGDEP      1.5674     0.1975    7.94   0.000

S = 0.326649    R-Sq = 74.4%    R-Sq(adj) - 73.8%

Analysis of Variance

Source           DF       SS       MS       F       P
Regression        2   30.012   15.006  140.64   0.000
Residual Error   97   10.350    0.107
Total            99   40.361

Source     DF   Seq SS
LOGSIZE     1   23.293
LOGDEP      1    6.718


Unusual Observations

Obs   LOGSIZE   LOGVALUE      Fit   SE Fit   Residual   St Resid
  3      7.40     9.7280  10.6725   0.0505    -0.9445    -2.93R
 36      7.54    10.5361  10.2537   0.1186     0.2824     0.93 X
 46      7.24    10.2454  10.9223   0.0488    -0.6769    -2.10R
 47      6.96     9.9496  10.6439   0.0632    -0.6944    -2.17R
 48      7.23     9.7898  10.6480   0.0342    -0.8582    -2.64R
 62      7.06    10.0623  10.8622   0.0685    -0.7998    -2.50R
 76      7.72    12.3202  11.6343   0.0759     0.6860     2.16R
 77      7.03    12.1118  10.4453   0.0373     1.6665     5.14R
 78      7.87    12.2140  12.0729   0.1016     0.1411     0.45 X
 97      8.18    12.3995  12.2065   0.1136     0.1931     0.63 X
 98      8.38    12.5990  12.4022   0.1313     0.1968     0.66 X

R denotes an observation with a large standardized residual.
X denotes an observation whose X value gives it large influence.
```

Note: To save space, the residual plots from the final model have not been included. However, to complete the analysis the following plots should be done: standardized residuals versus fitted values, standardized residuals versus each explanatory variable in the model, and standardized residuals versus each possible explanatory variable excluded from the model. There is one very large standardized residual that shows up in these plots (observation 77) with a standardized value of 5.14. Checking the numbers on this house shows that it is listed at 1126 square feet with a value of $182,012. Although I had no way of finding out if the numbers were correct, my guess is that the house should have been listed at 2126 square feet. This puts the value much more in line with the other data. Try this change and see what happens.

6.19 The regression using ACCIDENT as the dependent variable and VOLUME as the explanatory variable is shown. The residual plots from the regression are also shown as well as time series plots of Cook's D and Dfits. VOLUME is important in explaining the number of accidents, although it explains only 16.4% of the variation in y. From MINITAB's table of unusual observations (as well as from the residual plots), two observations stand out as unusual in the y-direction. These are observations 30 and 31. Both of these intersections have more accidents than would be expected given their volume of traffic. The city may want to investigate whether measures could be taken to make these intersections safer (install lights, remove obstructions, etc.). (Having lived in Fort Worth for over 20 years, I quickly recognized both of these intersections as being likely candidates for a high number of accidents. Both of these intersections have been redesigned and improved since the time this original data was collected.)

Regression of ACCIDENT on VOLUME

```
The regression equation is
ACCIDENT = 0.39 + 0.00200 VOLUME


Predictor        Coef     SE Coef        T        P
Constant        0.393       1.572     0.25    0.803
VOLUME       0.0019957   0.0005824     3.43    0.001


S = 4.44127    R-Sq = 16.4%    R-Sq(adj) = 15.0%


Analysis of Variance

Source            DF        SS         MS        F        P
Regression         1     231.60     231.60    11.74    0.001
Residual Error    60    1183.49      19.72
Total             61    1415.10


Unusual Observations

Obs   VOLUME   ACCIDENT     Fit    SE Fit   Residual   St Resid
  9     5811      7.000   11.990    1.999     -4.990     -1.26 X
 30     4250     28.000    8.875    1.155     19.125      4.46R
 31     2109     21.000    4.602    0.612     16.398      3.73R
```

R denotes an observation with a large standardized residual.
X denotes an observation whose X value gives it large influence.

Residual Plots from Initial Regression

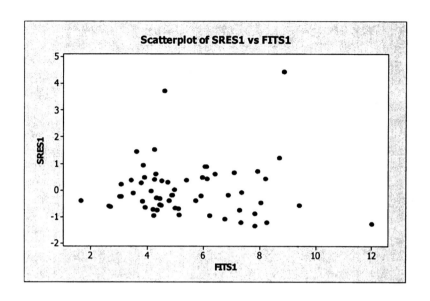

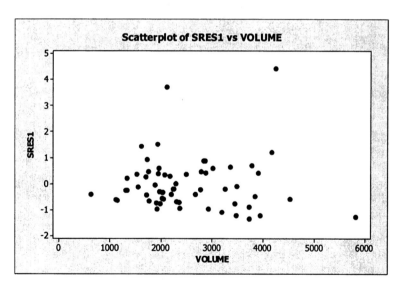

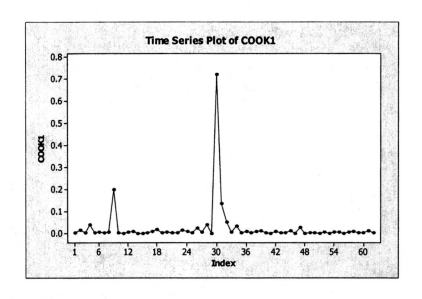

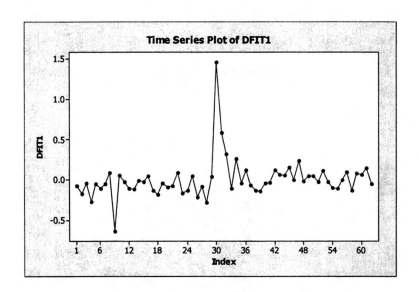

CHAPTER 7
Using Indicator and Interaction Variables

7.1 **a**   Hypotheses: $H_0: \beta_1 = \beta_2 = \beta_3 = \beta_4 = 0.0$
   H_a: At least one of the coefficients is not equal to 0.

   Decision Rule: Reject H_0 if $F > 2.53$ $F(0.05; 4, 88) \approx 2.53$
   Do not reject H_0 if $F \leq 2.53$

   Test Statistic: $F = 22.98$

   OR

   Decision Rule: Reject H_0 if p value < 0.05
   Do not reject H_0 if p value ≥ 0.05

   Test Statistic: p value $= 0.000$

   Decision: Reject H_0

b   At least one of the coefficients ($\beta_1, \beta_2, \beta_3, \beta_4$) is not equal to 0. At least one of the four explanatory variables (x_1, x_2, x_3, x_4) is important in explaining the variation in SALARY.

c   Hypotheses: $H_0: \beta_4 = 0.0$
   $H_a: \beta_4 \neq 0.0$

   Decision Rule: Reject H_0 if $t > 1.96$ or $t < -1.96$
   Do not reject H_0 if $-1.96 \leq t \leq 1.96$

   $t(0.025, 88) \approx 1.96$ (z value)

   Test Statistic: $t = 6.13$

   OR

   Decision Rule: Reject H_0 if p value < 0.05
   Do not reject H_0 if p value ≥ 0.05

   Test Statistic: p value $= 0.000$

   Decision: Reject H_0

d   Yes. There is a difference in salaries, on average, for male and female workers after accounting for the effects of the EDUC, EXPER, and MONTHS variables. Males' salaries are, on average, $722 higher, a statistically significant difference.

e   Forecast of average salary for males with 12 years education, 10 years of experience and with MONTHS equal to 15: $\hat{y}$ = 3526.422 + 722.461 + 90.020(12) + 1.269(10) + 23.406(15) = 5692.903

Forecast of average salary for females with 12 years education, 10 years of experience and with MONTHS equal to 15: $\hat{y}$ = 3526.422 + 90.020(12) + 1.269(10) + 23.406(15) = 4970.422

7.3   a R^2_{ADJ} = 49.4% with interaction variable

R^2_{ADJ} = 48.9% without interaction variable

Although the full model has a higher adjusted R-square value, the difference is very small. It is unclear from the adjusted R-square alone whether the interaction variable is necessary. The hypothesis test in part b) results in the conclusion that it is not.

b   Hypotheses: H_0: β_5 = 0.0
                        H_a: β_5 ≠ 0.0

Decision Rule: Reject H_0 if t > 1.96 or t < -1.96
                     Do not reject H_0 if -1.96 ≤ t ≤ 1.96

$t(0.025, 87)$ ≈ 1.96 (z value)

Test Statistic: t = -1.42

OR

Decision Rule: Reject H_0 if p value < 0.05
                     Do not reject H_0 if p value ≥ 0.05

Test Statistic: p value = 0.160

Decision: Do not reject H_0

c   The interaction term is not important in this regression model.

d   From the test results, it appears that the interaction term is not useful in explaining the difference in average salaries.

7.5 **a** Hypotheses: $H_0: \beta_3 = \beta_4 = \ldots = \beta_{12} = \beta_{13} = 0.0$
H_a: At least one of the coefficients is not equal to 0.

Decision Rule: Reject H_0 if $F > 1.99$ $F(0.05; 11, 117) \approx 1.99$
Do not reject H_0 if $F \leq 1.99$

Test Statistic: $F = \dfrac{(4903894 - 1017007)/11}{8692} = 40.653$

Decision: Reject H_0

Conclusion: There is seasonal variation in furniture sales.

b Hypotheses: $H_0: \beta_1 = 0.0$
$H_a: \beta_1 \neq 0.0$

Decision Rule: Reject H_0 if $t > 1.96$ or $t < -1.96$
Do not reject H_0 if $-1.96 \leq t \leq 1.96$

$t(0.025, 117) \approx 1.96$ (z value)

Test Statistic: $t = 6.09$

OR

Decision Rule: Reject H_0 if p value < 0.05
Do not reject H_0 if p value ≥ 0.05

Test Statistic: p value $= 0.000$

Decision: Reject H_0

Conclusion: The trend variable is important

Hypotheses: $H_0: \beta_2 = 0.0$
$H_a: \beta_2 \neq 0.0$

Decision Rule: Reject H_0 if $t > 1.96$ or $t < -1.96$
Do not reject H_0 if $-1.96 \leq t \leq 1.96$

$t(0.025, 117) \approx 1.96$ (z value)

Test Statistic: $t = 5.84$

OR

Decision Rule: Reject H_0 if p value < 0.05
Do not reject H_0 if p value ≥ 0.05

Test Statistic: p value = 0.000

Decision: Reject H_0

Conclusion: The lagged variable is important.

c The full model will be used. The following forecasts are computed by hand.

Date	Forecast
1/2003	SALES = 1552.151 + 7.903(133) + 0.484(4678) − 643.717 = 4223.685
2/2003	SALES = 1552.151 + 7.903(134) + 0.484(4223.685) − 404.481 = 4250.936
3/2003	SALES = 1552.151 + 7.903(135) + 0.484(4250.936) − 80.369 = 4596.140
4/2003	SALES = 1552.151 + 7.903(136) + 0.484(4596.140) − 457.714 = 4393.777
5/2003	SALES = 1552.151 + 7.903(137) + 0.484(4393.777) − 178.202 = 4583.248
6/2003	SALES = 1552.151 + 7.903(138) + 0.484(4583.248) − 316.796 = 4544.261
7/2003	SALES = 1552.151 + 7.903(139) + 0.484(4544.261) − 278.410 = 4571.680
8/2003	SALES = 1552.151 + 7.903(140) + 0.484(4571.680) − 183.319 = 4687.945
9/2003	SALES = 1552.151 + 7.903(141) + 0.484(4687.945) − 357.504 = 4577.935
10/2003	SALES = 1552.151 + 7.903(142) + 0.484(4577.935) − 239.408 = 4650.690
11/2003	SALES = 1552.151 + 7.903(143) + 0.484(4650.690) + 6.826 = 4947.943
12/2003	SALES = 1552.151 + 7.903(144) + 0.484(4947.943) = 5084.987

7.7 The first step in this problem is to see whether either HOURS or MATERIAL or both are related to DAYS. The initial scatterplots and the residual plots from the regression of DAYS on HOURS and MATERIAL (all shown on the following pages) indicate that the linearity assumption has been violated. All three of the variables require a log transformation.

Because DAYS and MATERIAL both have some 0s (for which the log transformation is not applicable) these variables are first transformed by adding 1 to each value. This transformation will not affect the analysis in any way. Just keep in mind that 1 has been added to each value if using the model for forecasting. The log transformation is then performed. The scatterplots after the log transformation show a much nicer linear relationship, as suspected.

The regression using the log transformed variables is shown along with the residual plots. The residual plots look random, so the violation has been successfully corrected. The MATERIAL variable is not significant in the regression, so it will be dropped.

Indicator variables are created for the four building types and the regression is rerun. To test whether there is a difference in the time taken to complete work orders, on average, for different building types, the hypothesis test proceeds as follows:

Hypotheses: $H_0: \beta_2 = \beta_3 = \beta_4 = 0.0$
H_a: At least one of the coefficients is not equal to zero.

Decision Rule: Reject H_0 if $F > 2.76$ $F(0.05; 3, 67) \approx 2.76$
Do not reject H_0 if $F \leq 2.76$

Test Statistic: $F = \dfrac{(46.6076 - 45.4320)/3}{0.6781} = 0.58$

Decision: Do not reject H_0

Conclusion: All of the coefficients of the indicator variables are equal to 0. There is NO difference in the time taken to complete work orders, on average, for different building types (as amazing as that may seem to the faculty)!

Scatterplots of DAYS versus HOURS and MATERIAL

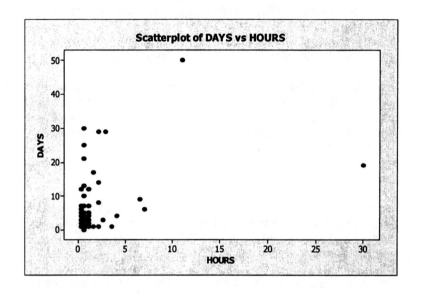

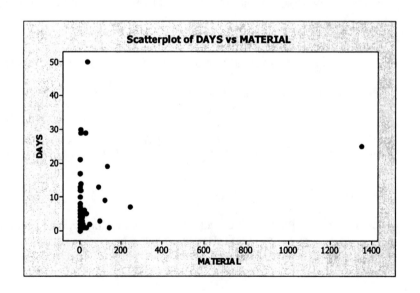

Regression of DAYS on HOURS and MATERIAL

```
The regression equation is
DAYS = 4.62 + 0.839 HOURS + 0.0137 MATERIAL

Predictor        Coef     SE Coef       T      P
Constant        4.625       1.028    4.50  0.000
HOURS          0.8389      0.2491    3.37  0.001
MATERIAL     0.013651    0.005842    2.34  0.022

S = 7.96770    R-Sq = 20.7%    R-Sq(adj) = 18.4%

Analysis of Variance

Source           DF        SS       MS      F      P
Regression        2   1140.46   570.23   8.98  0.000
Residual Error   69   4380.41    63.48
Total            71   5520.87

Source      DF    Seq SS
HOURS        1    793.83
MATERIAL     1    346.64

Unusual Observations

Obs   HOURS    DAYS      Fit    SE Fit   Residual   St Resid
 10    30.0  19.000   31.580     7.139    -12.580     -3.56RX
 22     2.8  29.000    6.986     1.007     22.014      2.79R
 34     2.0  29.000    6.644     0.948     22.356      2.83R
 39     0.5  25.000   23.473     7.764      1.527      0.85 X
 41     0.5  30.000    5.085     0.987     24.915      3.15R
 49    11.0  50.000   14.303     2.540     35.697      4.73R
 55     0.5  21.000    5.044     0.990     15.956      2.02R
```

R denotes an observation with a large standardized residual.
X denotes an observation whose X value gives it large influence.

Residual Plots from Initial Regression

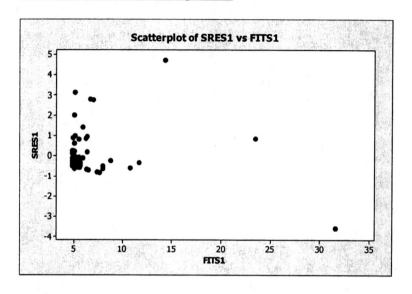

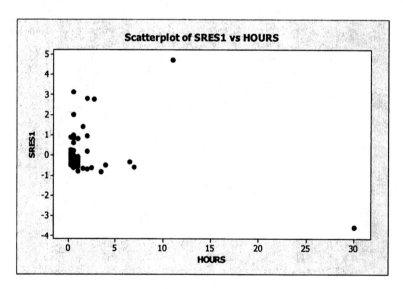

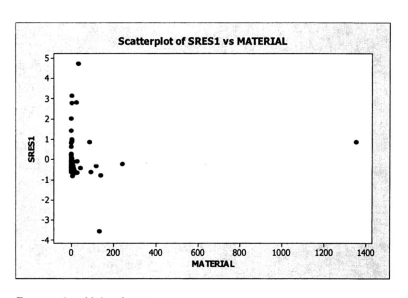

Regression Using Logs

```
The regression equation is
LOGDAYS = 1.58 + 0.313 LOGHOURS + 0.0496 LOGMAT

Predictor       Coef    SE Coef      T      P
Constant      1.5850     0.1473  10.76  0.000
LOGHOURS      0.3126     0.1144   2.73  0.008
LOGMAT       0.04956    0.06432   0.77  0.444

S = 0.818358   R-Sq = 15.7%   R-Sq(adj) = 13.2%

Analysis of Variance

Source            DF       SS      MS     F      P
Regression         2   8.5927  4.2963  6.42  0.003
Residual Error    69  46.2100  0.6697
Total             71  54.8027


Source      DF   Seq SS
LOGHOURS     1   8.1951
LOGMAT       1   0.3976

Unusual Observations

Obs   LOGHOURS  LOGDAYS     Fit   SE Fit  Residual  St Resid
 10       3.40   2.9957  2.8901   0.3895    0.1056     0.15 X
 39      -0.69   3.2581  1.7256   0.4076    1.5325     2.16RX
 41      -0.69   3.4340  1.4370   0.1054    1.9969     2.46R
 49       2.40   3.9318  2.5093   0.2915    1.4225     1.86 X
 55      -0.69   3.0910  1.3683   0.1246    1.7227     2.13R

R denotes an observation with a large standardized residual.
X denotes an observation whose X value gives it large influence.
```

Residual Plots from Regression Using Logs

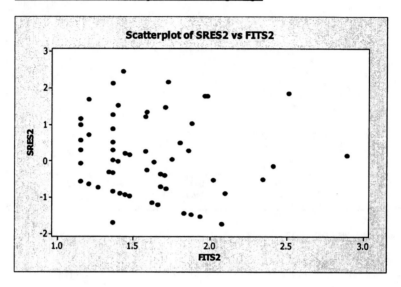

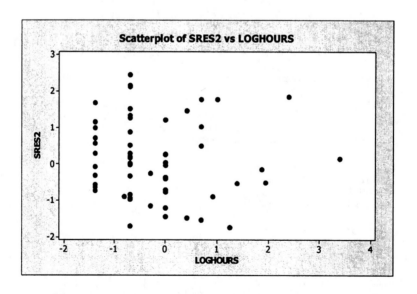

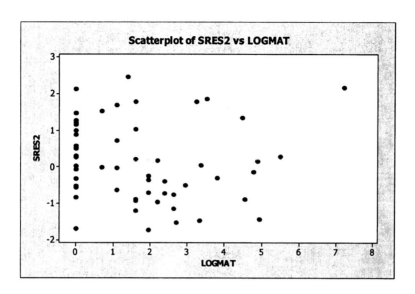

Regression Using Indicator Variables

```
The regression equation is
LOGDAYS = 1.57 + 0.352 LOGHOURS + 0.040 RESHALLS
          + 0.610 ATHLBLDG + 0.189 ACADBLDG

Predictor      Coef    SE Coef      T       P
Constant     1.5691     0.2657    5.91   0.000
LOGHOURS     0.3519     0.1055    3.34   0.001
RESHALLS     0.0396     0.2913    0.14   0.892
ATHLBLDG     0.6097     0.5439    1.12   0.266
ACADBLDG     0.1891     0.3271    0.58   0.565

S = 0.823462    R-Sq = 17.1%    R-Sq(adj) = 12.1%

Analysis of Variance

Source            DF        SS        MS       F       P
Regression         4    9.3707    2.3427    3.45   0.013
Residual Error    67   45.4320    0.6781
Total             71   54.8027

Source       DF    Seq SS
LOGHOURS      1    8.1951
RESHALLS      1    0.3016
ATHLBLDG      1    0.6472
ACADBLDG      1    0.2267


Unusual Observations

Obs   LOGHOURS   LOGDAYS      Fit    SE Fit   Residual   St Resid
 10       3.40    2.9957   2.9552    0.3998     0.0405       0.06 X
 28      -1.39    1.6094   1.6909    0.4779    -0.0815      -0.12 X
 39      -0.69    3.2581   1.9349    0.4761     1.3232       1.97 X
```

```
41    -0.69   3.4340  1.3648  0.1335   2.0692    2.55R
55    -0.69   3.0910  1.3648  0.1335   1.7263    2.12R
64    -0.69   0.6931  1.9349  0.4761  -1.2417   -1.85 X
```

R denotes an observation with a large standardized residual.
X denotes an observation whose X value gives it large influence.

Reduced Model Regression for Testing Differences in Building Types

```
The regression equation is
LOGDAYS = 1.67 + 0.354 LOGHOURS


Predictor      Coef    SE Coef        T       P
Constant     1.6670    0.1015     16.43   0.000
LOGHOURS     0.3538    0.1008      3.51   0.001


S = 0.815980    R-Sq = 15.0%    R-Sq(adj) = 13.7%


Analysis of Variance

Source            DF        SS       MS       F       P
Regression         1    8.1951   8.1951   12.31   0.001
Residual Error    70   46.6076   0.6658
Total             71   54.8027


Unusual Observations

Obs   LOGHOURS  LOGDAYS     Fit   SE Fit  Residual  St Resid
  2       1.87   2.3026  2.3293   0.2411   -0.0267    -0.03 X
 10       3.40   2.9957  2.8703   0.3875    0.1254     0.17 X
 12       1.95   1.9459  2.3555   0.2480   -0.4096    -0.53 X
 39      -0.69   3.2581  1.4218   0.1032    1.8363     2.27R
 41      -0.69   3.4340  1.4218   0.1032    2.0122     2.49R
 49       2.40   3.9318  2.5154   0.2906    1.4164     1.86 X
 55      -0.69   3.0910  1.4218   0.1032    1.6692     2.06R
```

R denotes an observation with a large standardized residual.
X denotes an observation whose X value gives it large influence.

Note: To save space, the residual plots from this model have not been included. However, to insure that no assumptions have been violated the following plots should be done: standardized residuals versus fitted values, standardized residuals versus each explanatory variable in the model, and standardized residuals versus each possible explanatory variable excluded from the model. Examining these plots reveals no further assumptions violated.

7.9 The initial regression includes a linear trend term and seasonal indicator variables. Before testing to determine whether there is seasonal variation or a linear trend, the regression will be examined for violations of assumptions. The Durbin-Watson statistic is used to test for first-order autocorrelation of the disturbances.

Hypotheses: $H_0: \rho = 0.0$
$H_a: \rho > 0.0$

Decision Rule: Reject H_0 if $d < 1.643$
Do not reject H_0 if $d > 1.896$
Inconclusive if $1.643 \leq d \leq 1.896$

(Critical values (approximate) from extended DW table in "The Durbin-Watson test for serial correlation with extreme sample sizes or many regressions," by N.E. Savin and K.J. White. *Econometrica* 45, 1977, pp. 1989-1996.)

Test Statistic: $d = 1.49519$

Decision: Reject H_0; there is autocorrelation

To correct for autocorrelation a lagged variable will be added to the regression. To test for first-order autocorrelation, Durbin's *h* test will be used:

Hypotheses: $H_0: \rho = 0.0$
$H_a: \rho > 0.0$

Decision Rule: Reject H_0 if $h > 1.645$
Do not reject H_0 if $h \leq 1.645$

Test Statistic: $r = 1 - 0.5d = 1 - 0.5(2.02256) = -0.01128$

$$h = (-0.01128)\sqrt{\frac{299}{1-(299)(0.05639)^2}} = -0.879$$

Decision: Do not reject H_0

Conclusion: First-order autocorrelation has been corrected.

In the regression with the trend, lagged variable, and the seasonal indicators, both the trend variable and the lagged variable are significant using a 5% level of significance, so both will be retained in the model. The hypothesis tests used are as follows:

Hypotheses: $H_0: \beta_1 = 0.0$
$H_a: \beta_1 \neq 0.0$

Decision Rule: Reject H_0 if $t > 1.96$ or $t < -1.96$
Do not reject H_0 if $-1.96 \leq t \leq 1.96$

$t(0.025, 285) \approx 1.96$ (z value)

Test Statistic: $t = 2.23$

OR

Decision Rule: Reject H_0 if p value < 0.05
Do not reject H_0 if p value ≥ 0.05

Test Statistic: p value $= 0.026$

Decision: Reject H_0

Conclusion: The trend variable is important.

Hypotheses: H_0: $\beta_{13} = 0.0$
H_a: $\beta_{13} \neq 0.0$

Decision Rule: Reject H_0 if $t > 1.96$ or $t < -1.96$
Do not reject H_0 if $-1.96 \leq t \leq 1.96$

$t(0.025, 285) \approx 1.96$ (z value)

Test Statistic: $t = 4.15$

OR

Decision Rule: Reject H_0 if p value < 0.05
Do not reject H_0 if p value ≥ 0.05

Test Statistic: p value $= 0.000$

Decision: Reject H_0

Conclusion: The lagged variable is important.

To test for seasonal effects, the reduced model regression with only the trend and lagged variable is needed. The two regressions being compared are the regression with the trend, lagged variable, and seasonal indicators (full model) and the regression with just the trend and lagged variable (reduced model). The hypotheses tested are as follows:

Hypotheses: H_0: $\beta_2 = \beta_3 = \ldots = \beta_{11} = \beta_{12} = 0.0$
H_a: At least one of the coefficients is not equal to 0.

Decision Rule: Reject H_0 if $F > 1.83$ $F(0.05; 11, 285) \approx 1.83$
Do not reject H_0 if $F \leq 1.83$

Test Statistic: $F = \dfrac{(19094 - 2370.2)/11}{8.3} = 183.17$

Decision: Reject H_0

Conclusion: There is seasonal variation in DFW temperatures (surprise!).

The forecasts shown here have been computed by hand using the full model.

Date   Forecast
1/03   33.2 + 0.00439(301) + 0.222 + 0.234(47.7) = 45.9
2/03   33.2 + 0.00439(302) + 4.89 + 0.234(45.9) = 50.2
3/03   33.2 + 0.00439(303) + 11.4 + 0.234(50.2) = 57.7
4/03   33.2 + 0.00439(304) + 18.1 + 0.234(57.7) = 66.1
5/03   33.2 + 0.00439(305) + 24.3 + 0.234(66.1) = 74.3
6/03   33.2 + 0.00439(306) + 30.1 + 0.234(74.3) = 82.0
7/03   33.2 + 0.00439(307) + 32.8 + 0.234(82.0) = 86.5
8/03   33.2 + 0.00439(308) + 31.4 + 0.234(86.5) = 86.2
9/03   33.2 + 0.00439(309) + 24.2 + 0.234(86.2) = 78.9
10/03   33.2 + 0.00439(310) + 15.2 + 0.234(78.9) = 68.2
11/03   33.2 + 0.00439(311) + 5.99 + 0.234(68.2) = 56.5
12/03   33.2 + 0.00439(312) + 0.234(56.5) = 47.8
1/04   33.2 + 0.00439(313) + 0.222 + 0.234(47.8) = 46.0
2/04   33.2 + 0.00439(314) + 4.89 + 0.234(46.0) = 50.2
3/04   33.2 + 0.00439(315) + 11.4 + 0.234(50.2) = 57.7
4/04   33.2 + 0.00439(316) + 18.1 + 0.234(57.7) = 66.2
5/04   33.2 + 0.00439(317) + 24.3 + 0.234(66.2) = 74.4
6/04   33.2 + 0.00439(318) + 30.1 + 0.234(74.4) = 82.1
7/04   33.2 + 0.00439(319) + 32.8 + 0.234(82.1) = 86.6
8/04   33.2 + 0.00439(320) + 31.4 + 0.234(86.6) = 86.3
9/04   33.2 + 0.00439(321) + 24.2 + 0.234(86.3) = 79.0
10/04   33.2 + 0.00439(322) + 15.2 + 0.234(79.0) = 68.3
11/04   33.2 + 0.00439(323) + 5.99 + 0.234(68.3) = 56.6
12/04   33.2 + 0.00439(324) + 0.234(56.6) = 47.9

Regression with Trend and Seasonal Indicators

```
The regression equation is
TEMP = 45.9 + 0.00660 TREND - 2.22 JAN + 2.34 FEB + 9.88 MARCH
       + 18.3 APRIL + 26.5 MAY + 34.3 JUNE + 38.8 JULY
       + 38.4 AUG + 31.1 SEPT + 20.4 OCT + 8.72 NOV
```

Predictor	Coef	SE Coef	T	P
Constant	45.8738	0.6806	67.40	0.000
TREND	0.006604	0.002014	3.28	0.001
JAN	-2.2234	0.8541	-2.60	0.010
FEB	2.3380	0.8541	2.74	0.007
MARCH	9.8754	0.8540	11.56	0.000
APRIL	18.3288	0.8540	21.46	0.000
MAY	26.5102	0.8539	31.04	0.000
JUNE	34.2796	0.8539	40.14	0.000
JULY	38.7770	0.8539	45.41	0.000
AUG	38.3944	0.8539	44.97	0.000
SEPT	31.1078	0.8539	36.43	0.000
OCT	20.4212	0.8538	23.92	0.000
NOV	8.7226	0.8538	10.22	0.000

S = 3.01875   R-Sq = 96.0%   R-Sq(adj) = 95.8%

Analysis of Variance

Source	DF	SS	MS	F	P
Regression	12	62240.2	5186.7	569.16	0.000
Residual Error	287	2615.4	9.1		
Total	299	64855.6			

Source	DF	Seq SS
TREND	1	145.7
JAN	1	12138.6
FEB	1	9371.8
MARCH	1	4531.8
APRIL	1	926.4
MAY	1	85.2
JUNE	1	2769.9
JULY	1	7555.5
AUG	1	10885.2
SEPT	1	8580.6
OCT	1	4298.6
NOV	1	951.0

Unusual Observations

Obs	TREND	TEMP	Fit	SE Fit	Residual	St Resid
1	1	33.800	43.657	0.670	-9.857	-3.35R
2	2	36.700	48.225	0.670	-11.525	-3.92R
13	13	35.400	43.736	0.660	-8.336	-2.83R
14	14	42.200	48.304	0.660	-6.104	-2.07R
30	30	87.000	80.352	0.650	6.648	2.26R
31	31	92.000	84.856	0.650	7.144	2.42R
72	72	34.800	46.349	0.627	-11.549	-3.91R
84	84	52.600	46.429	0.621	6.171	2.09R
85	85	37.800	44.212	0.616	-6.412	-2.17R

```
134    134    42.200    49.097    0.604    -6.897    -2.33R
144    144    39.000    46.825    0.604    -7.825    -2.65R
145    145    51.800    44.608    0.604     7.192     2.43R
221    221    80.800    73.843    0.621     6.957     2.35R
263    263    62.900    56.333    0.642     6.567     2.23R
266    266    57.300    49.968    0.650     7.332     2.49R
275    275    49.800    56.412    0.650    -6.612    -2.24R
276    276    39.400    47.696    0.650    -8.296    -2.81R
```

R denotes an observation with a large standardized residual.
Durbin-Watson statistic = 1.49519

Regression with Trend, Lagged Variable and Seasonal Indicators

The regression equation is
TEMP = 33.2 + 0.00439 TREND + 0.222 JAN + 4.89 FEB + 11.4 MARCH
       + 18.1 APRIL + 24.3 MAY + 30.1 JUNE + 32.8 JULY
       + 31.4 AUG + 24.2 SEPT + 15.2 OCT + 5.99 NOV
       + 0.234 TEMPL1

299 cases used, 1 case contains missing values

```
Predictor      Coef     SE Coef      T       P
Constant     33.214      3.146    10.56   0.000
TREND       0.004393   0.001969    2.23   0.026
JAN          0.2219     0.9596     0.23   0.817
FEB          4.891      1.023      4.78   0.000
MARCH       11.3622     0.8918    12.74   0.000
APRIL       18.0540     0.8184    22.06   0.000
MAY         24.2595     0.9793    24.77   0.000
JUNE        30.117      1.293     23.29   0.000
JULY        32.798      1.656     19.81   0.000
AUG         31.365      1.881     16.68   0.000
SEPT        24.168      1.861     12.98   0.000
OCT         15.186      1.503     10.10   0.000
NOV          5.987      1.049      5.71   0.000
TEMPL1       0.23381    0.05639    4.15   0.000
```

S = 2.88385   R-Sq = 96.3%   R-Sq(adj) = 96.1%

Analysis of Variance

```
Source            DF        SS        MS        F       P
Regression        13    61461.5    4727.8    568.48   0.000
Residual Error   285     2370.2       8.3
Total            298    63831.7
```

```
Source    DF    Seq SS
TREND      1      79.6
JAN        1   11282.9
FEB        1    9375.2
MARCH      1    4534.0
APRIL      1     927.3
MAY        1      84.9
JUNE       1    2768.8
JULY       1    7553.9
```

```
AUG      1   10883.6
SEPT     1    8579.6
OCT      1    4298.0
NOV      1     950.9
TEMPL1   1     143.0
```

```
Unusual Observations
Obs   TREND   TEMP     Fit   SE Fit  Residual  St Resid
  2       2  36.700  46.016   0.848    -9.316   -3.38R
 13      13  35.400  44.272   0.646    -8.872   -3.16R
 30      30  87.000  80.998   0.637     6.002    2.13R
 72      72  34.800  46.928   0.612   -12.128   -4.30R
 84      84  52.600  46.349   0.594     6.251    2.22R
 85      85  37.800  46.108   0.695    -8.308   -2.97R
134     134  42.200  50.384   0.655    -8.184   -2.91R
144     144  39.000  47.455   0.596    -8.455   -3.00R
145     145  51.800  43.192   0.736     8.608    3.09R
221     221  80.800  73.455   0.599     7.345    2.60R
263     263  62.900  56.536   0.617     6.364    2.26R
266     266  57.300  51.104   0.687     6.196    2.21R
275     275  49.800  56.705   0.628    -6.905   -2.45R
276     276  39.400  46.070   0.725    -6.670   -2.39R
279     279  51.800  57.539   0.631    -5.739   -2.04R
```

Durbin-Watson statistic = 2.02256

Residual Plots for Regression with Trend, Lagged Variable and Seasonal Indicators

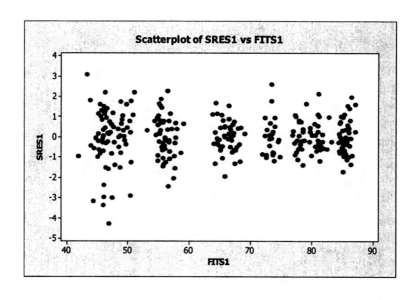

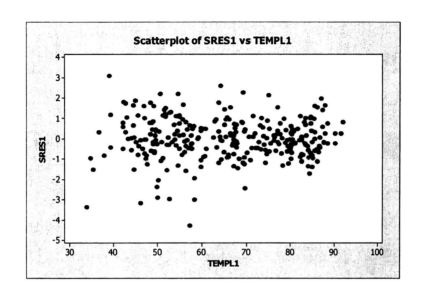

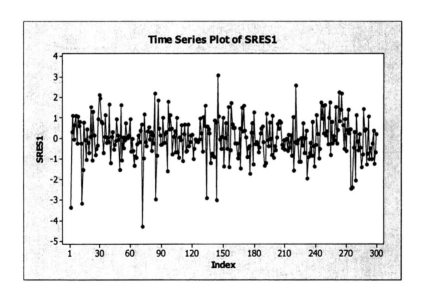

Reduced Model Regression with Lagged Variable Only

The regression equation is
TEMP = 11.3 - 0.00181 TREND + 0.833 TEMPL1

299 cases used, 1 case contains missing values

```
Predictor        Coef    SE Coef       T      P
Constant       11.299      2.244    5.04  0.000
TREND       -0.001807   0.005389   -0.34  0.738
TEMPL1        0.83316    0.03167   26.31  0.000


S = 8.03165    R-Sq = 70.1%    R-Sq(adj) = 69.9%
```

Analysis of Variance

```
Source            DF       SS      MS       F       P
Regression         2    44737   22369  346.76   0.000
Residual Error   296    19094      65
Total            298    63832


Source   DF   Seq SS
TREND     1       80
TEMPL1    1    44658
```

Unusual Observations

```
Obs  TREND    TEMP     Fit  SE Fit  Residual  St Resid
 23     23  52.900  70.245   0.854   -17.345    -2.17R
 72     72  34.800  58.908   0.676   -24.108    -3.01R
 85     85  37.800  54.969   0.708   -17.169    -2.15R
144    144  39.000  59.528   0.524   -20.528    -2.56R
156    156  44.000  60.839   0.503   -16.839    -2.10R
179    179  52.700  68.880   0.500   -16.180    -2.02R
221    221  80.800  64.388   0.602    16.412     2.05R
275    275  49.800  68.873   0.818   -19.073    -2.39R
```

R denotes an observation with a large standardized residual.

7.11 Following are time-series plots of FOC sales and log of FOC sales. The log transformation will be used in this problem. Neglecting this transformation will result in violation of the constant variance assumption. This becomes immediately evident from residual plots (not shown here).

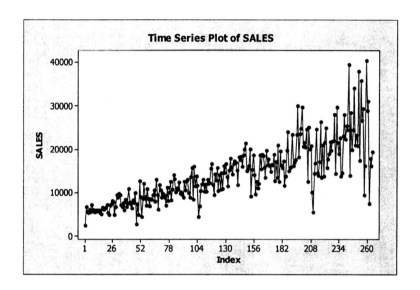

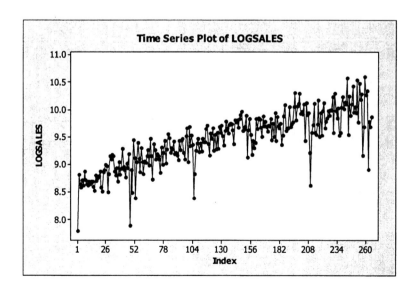

Regression of Log Sales on Linear Trend Variable

```
The regression equation is
LOGSALES = 8.74 + 0.00537 TREND

Predictor         Coef     SE Coef        T        P
Constant       8.73903     0.03430   254.80    0.000
TREND        0.0053746   0.0002235    24.04    0.000

S = 0.278373    R-Sq = 68.7%    R-Sq(adj) = 68.6%

Analysis of Variance

Source           DF        SS       MS        F        P
Regression        1    44.797   44.797   578.09    0.000
Residual Error  263    20.380    0.077
Total           264    65.177


Unusual Observations

Obs   TREND   LOGSALES     Fit   SE Fit   Residual   St Resid
  1       1     7.7936  8.7444   0.0341    -0.9508      -3.44R
 48      48     7.8966  8.9970   0.0256    -1.1005      -3.97R
 53      53     8.3891  9.0239   0.0247    -0.6348      -2.29R
105     105     8.3950  9.3034   0.0182    -0.9083      -3.27R
209     209     9.2183  9.8623   0.0241    -0.6440      -2.32R
210     210     8.6253  9.8677   0.0243    -1.2424      -4.48R
257     257     9.1668 10.1203   0.0326    -0.9535      -3.45R
262     262     8.9168 10.1472   0.0335    -1.2304      -4.45R

R denotes an observation with a large standardized residual.
```

Residual Plots

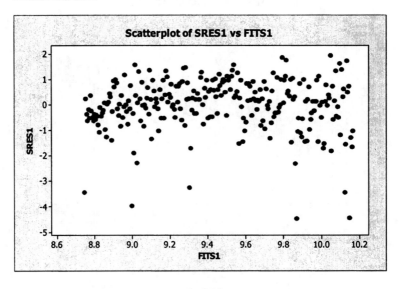

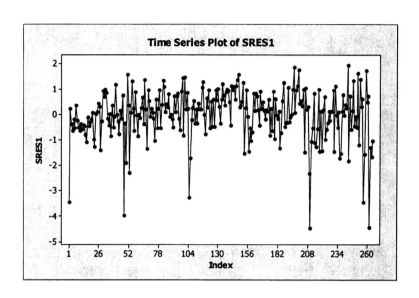

Note that there are several very large negative residuals. Most of these occur during the first few weeks of the year. Company officials suggested that this might be due to an end-of-year stockpiling of FOCs that resulted in a drop in sales at the first of the year. An indicator variable might be used to take account of this year-end effect.

7.13 Two additional variables were created to help with this analysis:

INDICATOR = 0 prior to November 1, 2002
= 1 on and after November 1, 2002

SLOPE = SP500*INDICATOR = 0 prior to November 1, 2002
= SP500 on and after November 1, 2002

INDICATOR allows for changes in the intercept of the line, SLOPE allows for changes in the slope. It is the change in the slope that is of interest here, but the intercept should also be allowed to change to provide flexibility in the fitting of the before and after November 1, 2002 regression lines. The regression of RET1 on SP500, INDICATOR, and SLOPE and the regression of RET2 on SP500, INDICATOR, and SLOPE are shown. Testing whether the slope of the regression lines has changed is performed by testing whether the coefficient of SLOPE is zero or not.

For RET1, the *p* value for the coefficient of SLOPE is 0.000 so we conclude that the coefficient is not equal to zero. The regression line prior to November 1, 2002 would be written as

RET1 = - 0.000146 + 1.09 SP500

while the regression line on and after November 1, 2002 would be

RET1 = - 0.000146 + 1.09 SP500 + 0.000222 INDICATOR - 0.0961 SLOPE

= 0.000076 + 0.9939 SP500

The beta changed from 1.09 to 0.9939 and the change was statistically significant.

For RET2, the *p* value for the coefficient of SLOPE is 0.000 so we conclude that the coefficient is not equal to zero. The regression line prior to November 1, 2002 would be written as

RET2 = - 0.000115 + 0.0796 SP500

while the regression line on and after November 1, 2002 would be

RET2 = - 0.000115 + 0.0796 SP500 + 0.000191 INDICATOR - 0.0509 SLOPE

= 0.000306 + 0.0287 SP500

The beta changed from 0.0796 to 0.0287 and the change was statistically significant. In both cases, the change in investment strategy resulted in a shift in the beta coefficient.

Regression of RET1 on SP500, INDICATOR, and SLOPE

The regression equation is
RET1 = - 0.000146 + 1.09 SP500 + 0.000222 INDICATOR
       - 0.0961 SLOPE

Predictor	Coef	SE Coef	T	P
Constant	-0.0001461	0.0001965	-0.74	0.458
SP500	1.08975	0.01157	94.20	0.000
INDICATOR	0.0002217	0.0002909	0.76	0.446
SLOPE	-0.09614	0.02026	-4.75	0.000

S = 0.00284089   R-Sq = 97.0%   R-Sq(adj) = 97.0%

Analysis of Variance

Source	DF	SS	MS	F	P
Regression	3	0.100880	0.033627	4166.53	0.000
Residual Error	382	0.003083	0.000008		
Total	385	0.103963			

Source	DF	Seq SS
SP500	1	0.100694
INDICATOR	1	0.000004
SLOPE	1	0.000182

Unusual Observations

Obs	SP500	RET1	Fit	SE Fit	Residual	St Resid
8	-0.0063	-0.013540	-0.006990	0.000205	-0.006550	-2.31R
10	-0.0163	-0.026250	-0.017855	0.000263	-0.008395	-2.97R
18	-0.0286	-0.041660	-0.031324	0.000374	-0.010336	-3.67R
34	-0.0155	-0.023250	-0.017048	0.000257	-0.006202	-2.19R
50	0.0114	0.006080	0.012255	0.000244	-0.006175	-2.18R
115	0.0009	0.008390	0.000878	0.000197	0.007512	2.65R
124	-0.0214	-0.017800	-0.023456	0.000306	0.005656	2.00R
131	0.0075	0.001160	0.008027	0.000220	-0.006867	-2.42R
132	-0.0065	-0.013720	-0.007175	0.000206	-0.006545	-2.31R
133	-0.0038	0.001590	-0.004244	0.000198	0.005834	2.06R
140	0.0573	0.067900	0.062307	0.000704	0.005593	2.03RX
143	0.0541	0.057300	0.058788	0.000668	-0.001488	-0.54 X
144	0.0043	0.010300	0.004485	0.000206	0.005815	2.05R
149	0.0299	0.038440	0.032459	0.000409	0.005981	2.13R
155	0.0401	0.041110	0.043498	0.000515	-0.002388	-0.85 X
168	-0.0415	-0.044450	-0.045414	0.000507	0.000964	0.34 X
170	-0.0160	-0.010000	-0.017528	0.000261	0.007528	2.66R
174	-0.0001	-0.006880	-0.000299	0.000196	-0.006581	-2.32R
175	-0.0248	-0.020130	-0.027150	0.000337	0.007020	2.49R
181	0.0025	-0.003650	0.002524	0.000200	-0.006174	-2.18R
182	-0.0138	-0.009430	-0.015217	0.000245	0.005787	2.04R
188	0.0400	0.048590	0.043466	0.000515	0.005124	1.83 X
196	0.0391	0.040970	0.042419	0.000504	-0.001449	-0.52 X
198	0.0473	0.049070	0.051443	0.000594	-0.002373	-0.85 X
229	0.0280	0.028710	0.027887	0.000501	0.000823	0.29 X
236	-0.0222	-0.028290	-0.021963	0.000437	-0.006327	-2.25R
249	-0.0160	-0.022760	-0.015852	0.000352	-0.006908	-2.45R

```
252    0.0332   0.034810    0.033064   0.000581    0.001746    0.63 X
267   -0.0292  -0.026330   -0.028968   0.000542    0.002638    0.95 X
272    0.0131   0.019840    0.013122   0.000297    0.006718    2.38R
300    0.0345   0.033680    0.034315   0.000600   -0.000635   -0.23 X
302    0.0354   0.034490    0.035279   0.000615   -0.000789   -0.28 X
306    0.0230   0.016380    0.022909   0.000427   -0.006529   -2.32R
307   -0.0352  -0.034390   -0.034929   0.000635    0.000539    0.19 X
```

R denotes an observation with a large standardized residual.
X denotes an observation whose X value gives it large influence.

Regression of RET2 on SP500, INDICATOR, and SLOPE

The regression equation is
RET2 = - 0.000115 + 0.0796 SP500 + 0.000191 INDICATOR - 0.0509 SLOPE

```
Predictor         Coef      SE Coef        T       P
Constant    -0.00011461   0.00009680    -1.18   0.237
SP500          0.079645     0.005699    13.97   0.000
INDICATOR      0.0001915    0.0001433    1.34   0.182
SLOPE         -0.050917     0.009981    -5.10   0.000
```

S = 0.00139965   R-Sq = 35.7%   R-Sq(adj) = 35.2%

Analysis of Variance

```
Source           DF          SS          MS       F       P
Regression        3   0.00041537  0.00013846   70.68   0.000
Residual Error  382   0.00074834  0.00000196
Total           385   0.00116371
```

```
Source       DF      Seq SS
SP500         1   0.00036113
INDICATOR     1   0.00000326
SLOPE         1   0.00005098
```

Unusual Observations

```
Obs    SP500      RET2        Fit      SE Fit   Residual  St Resid
  8  -0.0063  -0.003640  -0.000615   0.000101  -0.003025    -2.17R
 18  -0.0286  -0.005490  -0.002393   0.000184  -0.003097    -2.23R
 87   0.0375  -0.000720   0.002872   0.000240  -0.003592    -2.61R
124  -0.0214   0.001700  -0.001818   0.000151   0.003518     2.53R
131   0.0075  -0.006990   0.000483   0.000108  -0.007473    -5.36R
133  -0.0038  -0.003750  -0.000414   0.000098  -0.003336    -2.39R
140   0.0573   0.008300   0.004450   0.000347   0.003850     2.84RX
143   0.0541   0.008600   0.004193   0.000329   0.004407     3.24RX
144  -0.0043  -0.002700   0.000224   0.000101  -0.002924    -2.09R
151   0.0327   0.005370   0.002491   0.000216   0.002879     2.08R
155   0.0401   0.004580   0.003075   0.000254   0.001505     1.09 X
```

```
168  -0.0415  -0.006310  -0.003423   0.000250  -0.002887   -2.10RX
179  -0.0047   0.002310  -0.000485   0.000099   0.002795    2.00R
185   0.0182   0.004600   0.001336   0.000147   0.003264    2.35R
188   0.0400   0.000540   0.003073   0.000254  -0.002533   -1.84 X
195   0.0350  -0.000110   0.002671   0.000227  -0.002781   -2.01R
196   0.0391   0.000590   0.002996   0.000249  -0.002406   -1.75 X
198   0.0473  -0.001030   0.003656   0.000293  -0.004686   -3.42RX
200   0.0223  -0.001470   0.001661   0.000165  -0.003131   -2.25R
206   0.0172  -0.001990   0.001253   0.000142  -0.003243   -2.33R
229   0.0280   0.000570   0.000881   0.000247  -0.000311   -0.23 X
230  -0.0027   0.003280  -0.000002   0.000109   0.003282    2.35R
252   0.0332   0.002620   0.001031   0.000286   0.001589    1.16 X
267  -0.0292  -0.000070  -0.000763   0.000267   0.000693    0.50 X
281  -0.0016  -0.002770   0.000031   0.000107  -0.002801   -2.01R
300   0.0345  -0.000860   0.001067   0.000296  -0.001927   -1.41 X
302   0.0354   0.002990   0.001095   0.000303   0.001895    1.39 X
307  -0.0352  -0.001150  -0.000935   0.000313  -0.000215   -0.16 X
316   0.0027  -0.002920   0.000156   0.000107  -0.003076   -2.20R
349   0.0092   0.003820   0.000340   0.000126   0.003480    2.50R
```

R denotes an observation with a large standardized residual.
X denotes an observation whose X value gives it large influence.

7.15 Here is the process I followed in choosing which variables helped to explain total domestic gross revenue. This is not the only approach that could have been used here and I make no claims that this is the best model, but it is one systematic approach to try and find which variables have an effect on gross revenue.

After reviewing scatterplots, I decided to use the log of TDOMGROSS as the dependent variable. I called this variable LOGTDOM in the analyses that follow. I ran a regression with the two quantitative variables (BACTOR and TDACTOR) and the four indicator variables that had only two levels to represent (CHRISTMAS, HOLIDAY, SUMMER and SEQUEL). The variables used in these regressions were the ones that could be treated individually and evaluated for importance using t tests. This regression is labeled **Initial Regression.** I then performed a manual backward elimination, removing the weakest variables one at time.

Step 2 Regression: CHRISTMAS was the first variable removed (not shown)

Step 3 Regression: HOLIDAY was the second variable removed.

Step 4 Regression: The remaining variables were significant at the 5% level. Next, I evaluated the variables that needed to be treated as groups of indicators. There are three groups which we can call GENRE, MPAA and COUNTRY. The GENRE group consists of seven indicators used to represent the genre of the movie with one genre used as a base level. The base level was arbitrarily chosen to be science fiction. To evaluate the importance of GENRE, a partial F test is performed comparing the Step 4 Regression (Full Model) and the Step 3 Regression (Reduced Model). The resulting F statistic value is 10.3 which is significant at a 5% level (critical value about 2.01). So GENRE is important.

Step 5 Regression: The next group is MPAA (ratings). There are 5 ratings indicators used in this regression with unrated chosen (again arbitrarily) as the base level. The partial F test is performed comparing the Step 5 Regression (Full Model) and the Step 4 Regression (Reduced Model). The resulting F statistic value is 9.2 which is significant at a 5% level (critical value about 2.21). So GENRE is important.

Step 6 Regression: The final group is COUNTRY. There are 2 country indicators used in this regression with non-English speaking country chosen (again arbitrarily) as the base level. The partial F test is performed comparing the Step 5 Regression (Full Model) and the Step 4 Regression (Reduced Model). The resulting F statistic value is 1.1 which is not significant at a 5% level (critical value about 3.00). So COUNTRY is not important.

The final model chosen is the model in Step 5. The regression suggests that revenues increase with the number of "best actors" or "top dollar actors" in the movie. Summer releases add to revenue and so do sequels. The genre of the movie and the rating are also factors that affect revenue. Factors that were found to have no significant relationship to revenue were the country where the movie was made and Christmas or holiday release.

Initial Regression

```
LOGTDOM = - 0.343 + 1.41 BACTOR + 2.05 TDACTOR - 1.38 CHRISTMAS
              + 1.20 HOLIDAY + 0.871 SUMMER + 1.86 SEQUEL


Predictor        Coef     SE Coef        T        P
Constant      -0.3433      0.2068    -1.66    0.098
BACTOR         1.4127      0.2969     4.76    0.000
TDACTOR        2.0452      0.4140     4.94    0.000
CHRISTMAS     -1.3795      0.9491    -1.45    0.147
HOLIDAY        1.2030      0.6633     1.81    0.071
SUMMER         0.8710      0.3504     2.49    0.013
SEQUEL         1.8608      0.7614     2.44    0.015


S = 2.64488    R-Sq = 21.9%    R-Sq(adj) = 20.4%


Analysis of Variance

Source              DF          SS         MS         F        P
Regression           6     591.834     98.639     14.10    0.000
Residual Error     301    2105.612      6.995
Total              307    2697.446


Source         DF    Seq SS
BACTOR          1   255.861
TDACTOR         1   224.630
CHRISTMAS       1     4.615
HOLIDAY         1    28.398
SUMMER          1    36.544
SEQUEL          1    41.784


Unusual Observations

Obs   BACTOR   LOGTDOM       Fit    SE Fit   Residual   St Resid
  1     0.00    -1.177     1.517     0.746     -2.695      -1.06 X
  3     0.00     1.841    -0.520     0.710      2.361       0.93 X
  4     0.00     2.324     2.388     0.808     -0.065      -0.03 X
  9     1.00     0.025     0.893     0.692     -0.868      -0.34 X
 14     0.00     5.306     3.776     0.756      1.530       0.60 X
 17     0.00     2.908     2.720     0.969      0.187       0.08 X
 25     0.00    -7.960    -0.343     0.207     -7.617      -2.89 R
 29     0.00     2.638     1.517     0.746      1.120       0.44 X
 33     0.00     3.477     1.517     0.746      1.960       0.77 X
 35     1.00     5.092     2.272     0.698      2.820       1.11 X
```

44	1.00	-5.407	1.069	0.291	-6.476	-2.46R
63	1.00	-5.655	1.069	0.291	-6.724	-2.56R
72	1.00	1.709	0.893	0.692	0.816	0.32 X
73	0.00	-5.240	0.528	0.307	-5.767	-2.20R
74	0.00	-4.178	1.731	0.680	-5.909	-2.31R
78	0.00	3.690	-0.520	0.710	4.210	1.65 X
80	1.00	2.357	3.143	0.730	-0.786	-0.31 X
85	0.00	-5.175	0.528	0.307	-5.702	-2.17R
87	0.00	0.191	-0.520	0.710	0.711	0.28 X
96	0.00	-7.238	-0.343	0.207	-6.894	-2.61R
107	0.00	1.584	-0.520	0.710	2.104	0.83 X
109	0.00	-1.809	-0.520	0.710	-1.289	-0.51 X
110	0.00	-5.699	-0.343	0.207	-5.356	-2.03R
112	0.00	2.346	2.905	0.768	-0.559	-0.22 X
116	1.00	3.629	6.031	0.791	-2.402	-0.95 X
117	2.00	0.586	4.351	0.755	-3.765	-1.49 X
123	0.00	3.689	1.517	0.746	2.171	0.86 X
150	0.00	-1.924	1.517	0.746	-3.441	-1.36 X
152	0.00	4.865	8.524	1.344	-3.658	-1.61 X
162	0.00	1.267	1.517	0.746	-0.250	-0.10 X
169	0.00	3.925	1.525	0.762	2.399	0.95 X
187	0.00	2.938	1.517	0.746	1.421	0.56 X
194	0.00	-4.907	-0.520	0.710	-4.387	-1.72 X
197	1.00	4.905	2.938	0.698	1.967	0.77 X
205	1.00	1.375	0.893	0.692	0.482	0.19 X
212	2.00	3.663	6.572	0.839	-2.910	-1.16 X
213	2.00	4.617	6.396	0.924	-1.779	-0.72 X
225	0.00	-6.121	-0.343	0.207	-5.777	-2.19R
229	0.00	4.950	-0.343	0.207	5.293	2.01R
256	0.00	2.953	1.517	0.746	1.436	0.57 X
257	2.00	3.610	3.685	0.842	-0.075	-0.03 X
260	0.00	4.250	1.517	0.746	2.733	1.08 X
261	3.00	4.511	5.763	0.911	-1.252	-0.50 X
263	0.00	-5.785	-0.343	0.207	-5.442	-2.06R
274	1.00	-2.631	0.893	0.692	-3.524	-1.38 X
276	0.00	-6.229	-0.343	0.207	-5.886	-2.23R
280	1.00	4.057	4.975	0.811	-0.918	-0.36 X
292	0.00	4.385	2.905	0.768	1.480	0.58 X
307	1.00	4.751	2.938	0.698	1.813	0.71 X

R denotes an observation with a large standardized residual.
X denotes an observation whose X value gives it large influence.

Step 3 Regression

```
The regression equation is
LOGTDOM = - 0.299 + 1.39 BACTOR + 2.05 TDACTOR + 0.931 SUMMER
          + 1.90 SEQUEL


Predictor        Coef    SE Coef        T       P
Constant      -0.2985     0.2027    -1.47   0.142
BACTOR         1.3927     0.2937     4.74   0.000
TDACTOR        2.0488     0.4102     4.99   0.000
SUMMER         0.9307     0.3471     2.68   0.008
SEQUEL         1.8997     0.7616     2.49   0.013


S = 2.65099    R-Sq = 21.1%    R-Sq(adj) = 20.0%

Analysis of Variance

Source            DF        SS       MS       F       P
Regression         4    568.03   142.01   20.21   0.000
Residual Error   303   2129.41     7.03
Total            307   2697.45


Source     DF   Seq SS
BACTOR      1   255.86
TDACTOR     1   224.63
SUMMER      1    43.82
SEQUEL      1    43.72


Unusual Observations

Obs  BACTOR   LOGTDOM      Fit   SE Fit   Residual   St Resid
  1    0.00    -1.177    1.601    0.746     -2.779     -1.09 X
  4    0.00     2.324    2.532    0.805     -0.208     -0.08 X
 17    0.00     2.908    1.601    0.746      1.307      0.51 X
 25    0.00    -7.960   -0.299    0.203     -7.662     -2.90R
 29    0.00     2.638    1.601    0.746      1.036      0.41 X
 33    0.00     3.477    1.601    0.746      1.876      0.74 X
 44    1.00    -5.407    1.094    0.280     -6.501     -2.47R
 63    1.00    -5.655    1.094    0.280     -6.749     -2.56R
 73    0.00    -5.240    0.632    0.302     -5.872     -2.23R
 85    0.00    -5.175    0.632    0.302     -5.807     -2.20R
 96    0.00    -7.238   -0.299    0.203     -6.939     -2.63R
110    0.00    -5.699   -0.299    0.203     -5.401     -2.04R
116    1.00     3.629    6.122    0.773     -2.494     -0.98 X
123    0.00     3.689    1.601    0.746      2.087      0.82 X
150    0.00    -1.924    1.601    0.746     -3.525     -1.39 X
152    0.00     4.865    8.678    1.337     -3.813     -1.67 X
162    0.00     1.267    1.601    0.746     -0.334     -0.13 X
177    2.00     3.795    3.418    0.633      0.378      0.15 X
187    0.00     2.938    1.601    0.746      1.337      0.53 X
212    2.00     3.663    6.584    0.801     -2.922     -1.16 X
213    2.00     4.617    6.584    0.801     -1.967     -0.78 X
225    0.00    -6.121   -0.299    0.203     -5.822     -2.20R
```

```
256     0.00     2.953    1.601   0.746    1.352     0.53  X
260     0.00     4.250    1.601   0.746    2.649     1.04  X
261     3.00     4.511    5.928   0.778   -1.417    -0.56  X
263     0.00    -5.785   -0.299   0.203   -5.487    -2.08 R
276     0.00    -6.229   -0.299   0.203   -5.931    -2.24 R
280     1.00     4.057    5.043   0.806   -0.985    -0.39  X
283     0.00    -4.644    0.632   0.302   -5.277    -2.00 R
```

R denotes an observation with a large standardized residual.
X denotes an observation whose X value gives it large influence.

Step 4 Regression

The regression equation is
LOGTDOM = 2.35 + 1.08 BACTOR + 1.90 TDACTOR + 0.718 SUMMER
          + 0.458 SEQUEL - 0.731 ACTION - 3.06 DRAMA
          - 1.25 CHILDRENS - 2.38 COMEDY - 6.21 DOC
          - 1.71 THRILLER - 0.92 HORROR

```
Predictor        Coef    SE Coef       T      P
Constant       2.3535     0.8272    2.85  0.005
BACTOR         1.0789     0.2702    3.99  0.000
TDACTOR        1.8990     0.3809    4.99  0.000
SUMMER         0.7179     0.3180    2.26  0.025
SEQUEL         0.4584     0.7232    0.63  0.527
ACTION        -0.7305     0.9173   -0.80  0.426
DRAMA         -3.0642     0.8491   -3.61  0.000
CHILDRENS     -1.253      1.016    -1.23  0.218
COMEDY        -2.3781     0.8557   -2.78  0.006
DOC           -6.208      1.046    -5.94  0.000
THRILLER      -1.712      1.096    -1.56  0.119
HORROR        -0.919      1.045    -0.88  0.380
```

S = 2.40542   R-Sq = 36.5%   R-Sq(adj) = 34.1%

Analysis of Variance

```
Source           DF        SS       MS      F      P
Regression       11   984.772   89.525  15.47  0.000
Residual Error  296  1712.673    5.786
Total           307  2697.446
```

```
Source      DF    Seq SS
BACTOR       1   255.861
TDACTOR      1   224.630
SUMMER       1    43.818
SEQUEL       1    43.722
ACTION       1    85.678
DRAMA        1    34.791
CHILDRENS    1    17.016
COMEDY       1     0.886
DOC          1   264.254
```

```
THRILLER      1     9.641
HORROR        1     4.473
```

Unusual Observations

Obs	BACTOR	LOGTDOM	Fit	SE Fit	Residual	St Resid	
4	0.00	2.324	2.276	0.932	0.047	0.02	X
17	0.00	2.908	1.559	0.886	1.349	0.60	X
33	0.00	3.477	1.893	0.901	1.584	0.71	X
44	1.00	-5.407	0.368	0.302	-5.775	-2.42	R
51	0.00	-0.715	2.354	0.827	-3.068	-1.36	X
56	0.00	2.663	2.354	0.827	0.309	0.14	X
63	1.00	-5.655	0.368	0.302	-6.023	-2.52	
73	0.00	-5.240	0.693	0.335	-5.933	-2.49	R
74	0.00	-4.178	2.153	0.718	-6.331	-2.76	R
108	0.00	-4.834	-0.025	0.280	-4.809	-2.01	R
110	0.00	-5.699	-0.711	0.246	-4.989	-2.08	R
115	0.00	4.322	-0.711	0.246	5.033	2.10	R
123	0.00	3.689	1.893	0.901	1.795	0.80	X
152	0.00	4.865	8.496	1.251	-3.631	-1.77	X
158	1.00	4.236	3.432	0.832	0.803	0.36	X
213	2.00	4.617	7.056	0.891	-2.439	-1.09	X
218	0.00	-5.306	-0.025	0.280	-5.282	-2.21	R
225	0.00	-6.121	-0.711	0.246	-5.410	-2.26	R
249	0.00	2.679	2.354	0.827	0.325	0.14	X
253	0.00	-5.339	-0.025	0.280	-5.314	-2.22	R
256	0.00	2.953	2.812	0.975	0.141	0.06	X
257	2.00	3.610	4.511	0.921	-0.901	-0.41	X
259	0.00	1.948	2.354	0.827	-0.405	-0.18	X
260	0.00	4.250	2.812	0.975	1.438	0.65	X
263	0.00	-5.785	-0.711	0.246	-5.074	-2.12	R
268	0.00	-3.794	1.100	0.632	-4.894	-2.11	R
276	0.00	-6.229	-0.711	0.246	-5.519	-2.31	R
304	0.00	4.430	3.071	0.863	1.358	0.60	X

R denotes an observation with a large standardized residual.
X denotes an observation whose X value gives it large influence.

Step 5 Regression

The regression equation is
LOGTDOM = - 0.313 + 0.928 BACTOR + 1.59 TDACTOR + 0.760 SUMMER
         + 0.383 SEQUEL - 0.481 ACTION - 2.46 DRAMA
         - 1.60 CHILDRENS - 1.97 COMEDY - 4.38 DOC
         - 0.93 THRILLER - 0.428 HORROR + 3.57 G + 3.24 PG
         + 3.06 PG13 + 2.22 R + 2.13 NC17

Predictor	Coef	SE Coef	T	P
Constant	-0.3129	0.8802	-0.36	0.722
BACTOR	0.9277	0.2559	3.63	0.000
TDACTOR	1.5897	0.3625	4.39	0.000
SUMMER	0.7605	0.2996	2.54	0.012
SEQUEL	0.3828	0.6801	0.56	0.574
ACTION	-0.4814	0.8634	-0.56	0.578
DRAMA	-2.4586	0.8023	-3.06	0.002
CHILDRENS	-1.598	1.128	-1.42	0.158
COMEDY	-1.9686	0.8047	-2.45	0.015
DOC	-4.379	1.023	-4.28	0.000
THRILLER	-0.929	1.036	-0.90	0.371
HORROR	-0.4281	0.9940	-0.43	0.667
G	3.571	1.254	2.85	0.005
PG	3.2399	0.6033	5.37	0.000
PG13	3.0612	0.4864	6.29	0.000
R	2.2242	0.4325	5.14	0.000
NC17	2.126	1.646	1.29	0.197

S = 2.25348   R-Sq = 45.2%   R-Sq(adj) = 42.2%

Analysis of Variance

Source	DF	SS	MS	F	P
Regression	16	1219.703	76.231	15.01	0.000
Residual Error	291	1477.742	5.078		
Total	307	2697.446			

Source	DF	Seq SS
BACTOR	1	255.861
TDACTOR	1	224.630
SUMMER	1	43.818
SEQUEL	1	43.722
ACTION	1	85.678
DRAMA	1	34.791
CHILDRENS	1	17.016
COMEDY	1	0.886
DOC	1	264.254
THRILLER	1	9.641
HORROR	1	4.473
G	1	2.306
PG	1	28.133
PG13	1	70.005

```
R        1   126.015
NC17     1     8.472
```

Unusual Observations

Obs	BACTOR	LOGTDOM	Fit	SE Fit	Residual	St Resid
4	0.00	2.324	2.803	1.069	-0.479	-0.24 X
17	0.00	2.908	1.712	0.981	1.196	0.59 X
44	1.00	-5.407	0.380	0.306	-5.787	-2.59R
67	0.00	-3.388	-1.910	0.932	-1.478	-0.72 X
68	1.00	-3.395	1.396	0.551	-4.791	-2.19R
74	0.00	-4.178	2.244	0.673	-6.422	-2.99R
90	0.00	1.855	-2.771	0.419	4.626	2.09R
108	0.00	-4.834	-0.057	0.296	-4.776	-2.14R
110	0.00	-5.699	-0.547	0.257	-5.152	-2.30R
152	0.00	4.865	7.342	1.216	-2.477	-1.31 X
175	0.00	4.793	4.010	0.923	0.782	0.38 X
191	0.00	-0.541	-0.156	1.606	-0.386	-0.24 X
213	2.00	4.617	6.364	0.959	-1.747	-0.86 X
218	0.00	-5.306	-0.057	0.296	-5.249	-2.35R
225	0.00	-6.121	-0.547	0.257	-5.573	-2.49R
230	0.00	-3.853	0.703	0.339	-4.556	-2.05R
234	1.00	-4.094	1.217	0.376	-5.312	-2.39R
253	0.00	-5.339	-0.057	0.296	-5.282	-2.36R
256	0.00	2.953	2.294	0.940	0.659	0.32 X
260	0.00	4.250	3.310	0.981	0.940	0.46 X
263	0.00	-5.785	0.290	0.362	-6.075	-2.73R
268	0.00	-3.794	1.660	0.869	-5.454	-2.62R
270	1.00	-4.164	1.217	0.376	-5.381	-2.42R
275	0.00	5.173	0.703	0.339	4.470	2.01R
276	0.00	-6.229	-0.547	0.257	-5.682	-2.54R
279	1.00	0.668	0.282	1.606	0.386	0.24 X
297	0.00	-1.354	2.919	0.808	-4.273	-2.03R

R denotes an observation with a large standardized residual.
X denotes an observation whose X value gives it large influence.

Step 6 Regression

```
The regression equation is
LOGTDOM = - 0.463 + 0.883 BACTOR + 1.53 TDACTOR + 0.803 SUMMER
         + 0.367 SEQUEL - 0.550 ACTION - 2.45 DRAMA
         - 1.58 CHILDRENS - 2.04 COMEDY - 4.57 DOC
         - 0.98 THRILLER - 0.483 HORROR + 3.45 G + 3.07 PG
         + 2.94 PG13 + 2.08 R + 1.90 NC17 + 0.432 USA
         - 0.130 ENG
```

Predictor	Coef	SE Coef	T	P
Constant	-0.4631	0.9151	-0.51	0.613
BACTOR	0.8825	0.2579	3.42	0.001
TDACTOR	1.5329	0.3646	4.20	0.000
SUMMER	0.8030	0.3021	2.66	0.008
SEQUEL	0.3669	0.6800	0.54	0.590
ACTION	-0.5501	0.8664	-0.63	0.526
DRAMA	-2.4508	0.8046	-3.05	0.003
CHILDRENS	-1.580	1.129	-1.40	0.163
COMEDY	-2.0440	0.8060	-2.54	0.012
DOC	-4.568	1.031	-4.43	0.000
THRILLER	-0.984	1.042	-0.94	0.346
HORROR	-0.4825	0.9958	-0.48	0.628
G	3.450	1.283	2.69	0.008
PG	3.0692	0.6467	4.75	0.000
PG13	2.9377	0.5296	5.55	0.000
R	2.0795	0.4824	4.31	0.000
NC17	1.900	1.661	1.14	0.254
USA	0.4324	0.4729	0.91	0.361
ENG	-0.1295	0.6044	-0.21	0.830

S = 2.25254   R-Sq = 45.6%   R-Sq(adj) = 42.3%

Analysis of Variance

Source	DF	SS	MS	F	P
Regression	18	1231.075	68.393	13.48	0.000
Residual Error	289	1466.370	5.074		
Total	307	2697.446			

Source	DF	Seq SS
BACTOR	1	255.861
TDACTOR	1	224.630
SUMMER	1	43.818
SEQUEL	1	43.722
ACTION	1	85.678
DRAMA	1	34.791
CHILDRENS	1	17.016
COMEDY	1	0.886
DOC	1	264.254
THRILLER	1	9.641
HORROR	1	4.473

```
G         1    2.306
PG        1   28.133
PG13      1   70.005
R         1  126.015
NC17      1    8.472
USA       1   11.139
ENG       1    0.233
```

Unusual Observations

Obs	BACTOR	LOGTDOM	Fit	SE Fit	Residual	St Resid
4	0.00	2.324	2.448	1.111	-0.124	-0.06 X
17	0.00	2.908	1.825	0.983	1.082	0.53 X
44	1.00	-5.407	0.480	0.314	-5.887	-2.64R
68	1.00	-3.395	1.470	0.553	-4.865	-2.23R
74	0.00	-4.178	1.937	0.781	-6.115	-2.89R
108	0.00	-4.834	0.005	0.301	-4.838	-2.17R
110	0.00	-5.699	-0.402	0.275	-5.297	-2.37R
152	0.00	4.865	7.267	1.220	-2.402	-1.27 X
191	0.00	-0.541	-0.175	1.605	-0.367	-0.23 X
194	0.00	-4.907	-0.402	0.275	-4.505	-2.02R
218	0.00	-5.306	0.005	0.301	-5.311	-2.38R
225	0.00	-6.121	-0.402	0.275	-5.719	-2.56R
230	0.00	-3.853	0.808	0.346	-4.661	-2.09R
234	1.00	-4.094	1.339	0.385	-5.433	-2.45R
241	0.00	-2.083	2.357	0.496	-4.440	-2.02R
253	0.00	-5.339	0.005	0.301	-5.344	-2.39R
260	0.00	4.250	3.405	0.984	0.845	0.42 X
263	0.00	-5.785	0.456	0.379	-6.241	-2.81R
268	0.00	-3.794	1.278	0.918	-5.072	-2.47R
270	1.00	-4.164	1.339	0.385	-5.503	-2.48R
276	0.00	-6.229	-0.402	0.275	-5.827	-2.61R
279	1.00	0.668	0.301	1.605	0.367	0.23 X
297	0.00	-1.354	2.991	0.809	-4.345	-2.07R

R denotes an observation with a large standardized residual.
X denotes an observation whose X value gives it large influence.

7.17 The equation determined by the regression is

COST = 20.8 + 0.0122 PAGES - 10.8 SOFTCOVER.

Each additional page in a book adds 1.22 cents to the cost. Softcover books cost $10.80 less, on average, than hardcover books.

Regression of COST on PAGES and SOFTCOVER

```
 The regression equation is
 COST = 20.8 + 0.0122 PAGES - 10.8 SOFTCOVER

Predictor       Coef    SE Coef        T        P
Constant     20.7867     0.7385    28.15    0.000
PAGES       0.012164   0.002048     5.94    0.000
SOFTCOVER   -10.7986     0.6523   -16.55    0.000

S = 4.59902    R-Sq = 60.0%    R-Sq(adj) = 59.6%

Analysis of Variance

Source            DF        SS       MS        F        P
Regression         2    6481.2   3240.6   153.21    0.000
Residual Error   204    4314.8     21.2
Total            206   10796.0

Source         DF    Seq SS
PAGES           1     684.7
SOFTCOVER       1    5796.5

Unusual Observations

Obs   PAGES     COST     Fit   SE Fit   Residual   St Resid
  2     152   35.000  22.636    0.511     12.364      2.71R
  3      64   40.000  21.565    0.634     18.435      4.05R
  4      80   24.950  10.961    0.682     13.989      3.08R
 56     352   39.950  25.069    0.427     14.881      3.25R
 60    1200   35.000  35.384    1.891     -0.384     -0.09 X
 69     352   35.000  25.069    0.427      9.931      2.17R
 70     858   45.000  31.224    1.217     13.776      3.11RX
 71     832   16.000  20.109    1.194     -4.109     -0.93 X
 72     825   16.000  20.024    1.181     -4.024     -0.91 X
102      80   10.000  21.760    0.610    -11.760     -2.58R
103     912   18.000  21.082    1.344     -3.082     -0.70 X
108     224   22.500  12.713    0.531      9.787      2.14R
113     144   25.000  11.740    0.602     13.260      2.91R
124     224   35.000  23.512    0.441     11.488      2.51R
138     450   24.950  15.462    0.587      9.488      2.08R
156     240   35.000  23.706    0.430     11.294      2.47R
169     184   40.000  23.025    0.475     16.975      3.71R
170     192   40.000  23.122    0.467     16.878      3.69R
172     120   10.000  22.246    0.552    -12.246     -2.68R

R denotes an observation with a large standardized residual.
X denotes an observation whose X value gives it large influence.
```

CHAPTER 8
Variable Selection

8.1   **a** $\hat{y} = 59.43 + 0.95x_1 + 2.39x_2$ or

COST = 59.43PAPER + 2.39MACHINE

      b   99.87%

      c   0.9986 or 99.86%

      d $s_e = 11.0$

      e   OVERHEAD and LABOR

Overhead and labor are related to COST when examined individually. However, they add little to the ability of the equation to explain the variation in COST. The variables PAPER and MACHINE explain over 99% of the variation in COST. OVERHEAD and LABOR are probably unnecessary because most of the variation in COST is explained by the other two variables.

In addition to the backward elimination results, there is a regression of COST on all four variables showing the variance inflation factors and a correlation matrix showing the pairwise correlations between the four explanatory variables. Note the high VIFs and high pairwise correlations.

Backward Elimination Results

```
Backward elimination.   Alpha-to-Remove: 0.1

Response is COST on 4 predictors, with N = 27


Step                     1         2         3
Constant             51.72     51.17     59.43

PAPER                 0.95      0.94      0.95
T-Value               7.90      8.69      8.62
P-Value              0.000     0.000     0.000

MACHINE               2.47      2.51      2.39
T-Value               5.31     11.01     11.36
P-Value              0.000     0.000     0.000

OVERHEAD              0.05
T-Value               0.09
P-Value              0.927

LABOR               -0.051    -0.051
T-Value              -1.26     -1.29
P-Value              0.223     0.210

S                     11.1      10.8      11.0
R-Sq                 99.88     99.88     99.87
R-Sq(adj)            99.86     99.87     99.86
Mallows C-p            5.0       3.0       2.6
```

Regression Output with VIFs

```
The regression equation is
COST = 51.7 + 0.948 PAPER + 2.47 MACHINE + 0.048 OVERHEAD
            - 0.0506 LABOR


Predictor         Coef    SE Coef        T       P      VIF
Constant         51.72      21.70     2.38   0.026
PAPER           0.9479     0.1200     7.90   0.000     55.5
MACHINE         2.4710     0.4656     5.31   0.000    228.9
OVERHEAD        0.0483     0.5250     0.09   0.927    104.1
LABOR         -0.05058    0.04030    -1.26   0.223      9.3


S = 11.0756    R-Sq = 99.9%    R-Sq(adj) = 99.9%


Analysis of Variance

Source            DF        SS        MS         F       P
Regression         4   2271423    567856   4629.17   0.000
Residual Error    22      2699       123
Total             26   2274122


Source       DF    Seq SS
PAPER         1   2255666
MACHINE       1     15561
OVERHEAD      1         3
LABOR         1       193


Unusual Observations

Obs   PAPER     COST       Fit   SE Fit   Residual   St Resid
 17     891  1828.00   1823.22     8.68       4.78       0.69 X
 25     647  1317.00   1294.48     3.58      22.52       2.15R

R denotes an observation with a large standardized residual.
X denotes an observation whose X value gives it large influence.
```

Correlation Results

```
                 PAPER    MACHINE   OVERHEAD
MACHINE          0.989
                 0.000

OVERHEAD         0.978      0.994
                 0.000      0.000

LABOR            0.933      0.945      0.938
                 0.000      0.000      0.000

Cell Contents: Pearson correlation
               P-Value
```

8.3 I used mileage in city driving as my dependent variable. The inverse (reciprocal) transformation was used for each of the candidate independent variables: WTINV is the inverse of WEIGHT; HPINV is the inverse of HP; CYLINV is the inverse of CYLIN; LITINV is the inverse of LITER.

<u>Best Subsets Regression</u>

```
Response is CITYMPG
```

					W T I N V	H P I N V	C Y L I N V	L I T I N V
Vars	R-Sq	R-Sq(adj)	Mallows C-p	S				
1	78.9	78.8	30.7	3.1155		X		
1	72.8	72.6	81.4	3.5410				X
1	51.3	50.9	258.8	4.7387			X	
2	80.3	80.0	21.4	3.0231		X		X
2	79.4	79.1	29.2	3.0948		X	X	
2	79.2	78.9	30.7	3.1084	X	X		
3	82.8	82.4	3.1	2.8373		X	X	X
3	80.3	79.9	23.4	3.0337	X	X		X
3	79.7	79.3	28.1	3.0773	X	X	X	
4	82.8	82.3	5.0	2.8466	X	X	X	X

<u>Stepwise Regression</u>

```
   Alpha-to-Enter: 0.15  Alpha-to-Remove: 0.15

Response is CITYMPG on 4 predictors, with N = 147

Step                 1       2       3
Constant         5.890   5.219   7.609

HPINV             2769    1998    2235
T-Value          23.31    7.41    8.65
P-Value          0.000   0.000   0.000

LITINV                    12.9    24.6
T-Value                   3.16    5.32
P-Value                  0.002   0.000

CYLINV                           -42.1
T-Value                          -4.53
P-Value                          0.000

S                 3.12    3.02    2.84
R-Sq             78.94   80.31   82.78
R-Sq(adj)        78.80   80.04   82.41
Mallows C-p       30.7    21.4     3.1
```

Forward Selection

```
Forward selection.  Alpha-to-Enter: 0.25

Response is CITYMPG on 4 predictors, with N = 147

Step              1      2      3
Constant      5.890  5.219  7.609

HPINV          2769   1998   2235
T-Value       23.31   7.41   8.65
P-Value       0.000  0.000  0.000

LITINV                12.9   24.6
T-Value               3.16   5.32
P-Value              0.002  0.000

CYLINV                      -42.1
T-Value                     -4.53
P-Value                     0.000

S              3.12   3.02   2.84
R-Sq          78.94  80.31  82.78
R-Sq(adj)     78.80  80.04  82.41
Mallows C-p    30.7   21.4    3.1
```

Backward Elimination

```
Backward elimination.  Alpha-to-Remove: 0.1

Response is CITYMPG on 4 predictors, with N = 147

Step              1      2
Constant      7.981  7.609

WTINV         -1865
T-Value       -0.27
P-Value       0.786

HPINV          2246   2235
T-Value        8.56   8.65
P-Value       0.000  0.000

CYLINV        -42.3  -42.1
T-Value       -4.52  -4.53
P-Value       0.000  0.000

LITINV         25.1   24.6
T-Value        5.01   5.32
P-Value       0.000  0.000

S              2.85   2.84
R-Sq          82.78  82.78
R-Sq(adj)     82.30  82.41
Mallows C-p     5.0    3.1
```

Possible Final Regression

```
The regression equation is
CITYMPG = 5.22 + 1998 HPINV + 12.9 LITINV


Predictor     Coef    SE Coef      T       P
Constant    5.2191     0.7025    7.43   0.000
HPINV       1998.5      269.6    7.41   0.000
LITINV      12.943      4.093    3.16   0.002


S = 3.02312    R-Sq = 80.3%    R-Sq(adj) = 80.0%


Analysis of Variance

Source            DF       SS       MS       F       P
Regression         2   5367.3   2683.7  293.64   0.000
Residual Error   144   1316.1      9.1
Total            146   6683.4


Source   DF   Seq SS
HPINV     1   5275.9
LITINV    1     91.4

Unusual Observations

Obs    HPINV   CITYMPG     Fit   SE Fit   Residual   St Resid
 11   0.0056    22.000  23.512    0.750     -1.512      -0.52 X
 55   0.0118    47.000  38.687    0.779      8.313       2.85RX
 56   0.0149    61.000  47.990    1.152     13.010       4.66RX
 57   0.0042    20.000  20.017    0.867     -0.017      -0.01 X
 94   0.0053    20.000  22.983    0.811     -2.983      -1.02 X
104   0.0037    18.000  19.092    0.984     -1.092      -0.38 X
112   0.0031    22.000  14.278    0.371      7.722       2.57R
138   0.0102    52.000  34.240    0.609     17.760       6.00R

R denotes an observation with a large standardized residual.
X denotes an observation whose X value gives it large influence.
```

Residual Plots

Standardized Residuals versus Fitted Values

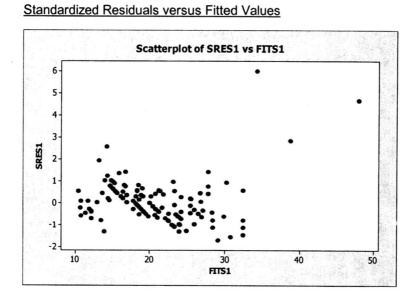

Standardized Residuals versus Variables in the Model

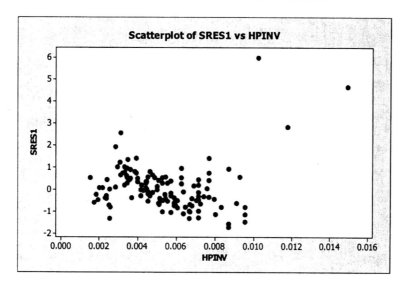

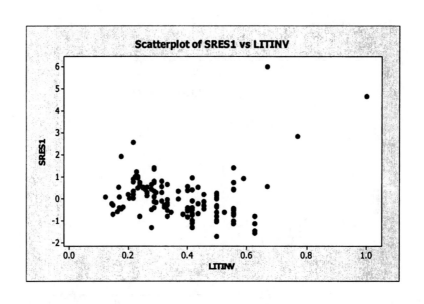

Standardized Residuals versus Variables Not in the Model

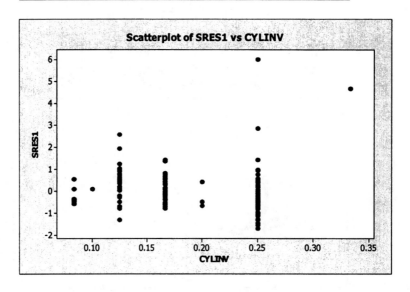

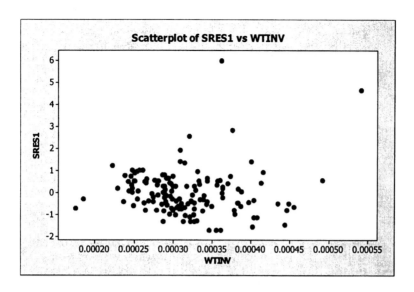

a   CITYMPG = 5.22 + 1998 HPINV + 12.9 LITINV

CITYMPG = mileage in city driving
HPINV = 1/HP = reciprocal of horsepower
LITINV = 1/LITER = reciprocal of liter

b   The reciprocal transformation was used on all variables in the equation. The scatterplots of CITYMPG versus each x variable suggest that this may provide a better fit. If the linear model is fit, the residual plots will also reveal the need for some sort of transformation to correct for nonlinearity. From the residual plots for the transformed model, the violation of the linearity assumption appears to be corrected.

There are some very large residuals that we might want to consider: Observation 56, the Honda Insight, has a standardized residual of 4.66. Observation 138, the Toyota Prius, has a standardized residual of 6.00. These are both hybrid cars. We might want to consider fitting the model without these two cars because the technology is different from most of the cars in the sample. Or we could use an indicator variable for hybrid cars to set them apart from the others. The Honda Civic Hybrid also falls in this category so let's create an indicator variable, HYBRID, that is one for these three cars and zero for all the others. Here is the resulting regression and the residual plots. The residual plots look better now because we have explained the presence of the extreme outliers.

Regression Including HYBRID Indicator Variable

```
The regression equation is
CITYMPG = 7.76 + 1641 HPINV + 10.4 LITINV + 17.0 HYBRID


Predictor      Coef   SE Coef       T      P
Constant     7.7632    0.5394   14.39  0.000
HPINV        1641.0     192.8    8.51  0.000
LITINV       10.362     2.900    3.57  0.000
HYBRID       16.976     1.408   12.06  0.000


S = 2.13623    R-Sq = 90.2%    R-Sq(adj) = 90.0%


Analysis of Variance

Source             DF       SS      MS       F      P
Regression          3   6030.8  2010.3  440.52  0.000
Residual Error    143    652.6     4.6
Total             146   6683.4


Source   DF   Seq SS
HPINV     1   5275.9
LITINV    1     91.4
HYBRID    1    663.5


Unusual Observations

Obs    HPINV   CITYMPG      Fit   SE Fit  Residual  St Resid
 27   0.0029    19.000   14.270    0.316     4.730     2.24R
 43   0.0025    10.000   14.796    0.372    -4.796    -2.28R
 44   0.0025    10.000   14.796    0.372    -4.796    -2.28R
 54   0.0087    33.000   28.128    0.369     4.872     2.32R
 55   0.0118    47.000   52.015    1.235    -5.015    -2.88RX
 56   0.0149    61.000   59.593    1.261     1.407     0.82 X
 57   0.0042    20.000   19.781    0.613     0.219     0.11 X
104   0.0037    18.000   19.022    0.695    -1.022    -0.51 X
112   0.0031    22.000   15.144    0.272     6.856     3.24R
134   0.0077    32.000   26.143    0.319     5.857     2.77R
138   0.0102    52.000   48.392    1.250     3.608     2.08RX

R denotes an observation with a large standardized residual.
X denotes an observation whose X value gives it large influence.
```

Residual Plots

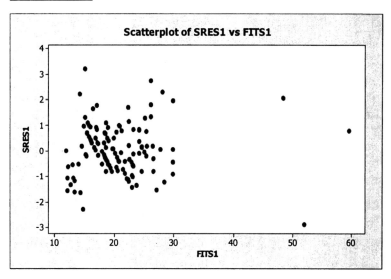

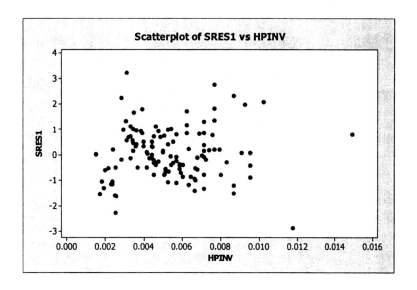

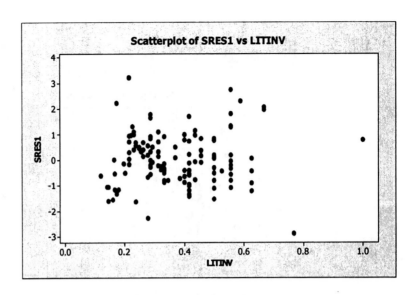

c   I choose the model with HPINV and LITINV and choose to omit CYLINV which showed up as significant statistically, but had a sign opposite to what I expected. I also chose to include an indicator variable for hybrid cars. Practically speaking, we have: As HP increases, mileage will drop, although not at a linear rate; as the engine size increases (LITERS), the mileage will also drop, although not at a linear rate. These relationships make sense. The interpretation of the equation provides some additional information as well. Hybrid cars will get 17 mpg more, on average, than cars with gasoline engines and the increase is statistically significant (p value = 0.000). This is a huge increase in mileage to say the least. (Also note that CYLINV is no longer significant when the hybrid variable is included. This model makes more sense to me than including CYLINV with a sign opposite to what should be expected.)

CHAPTER 9
An Introduction to Analysis of Variance

9.1 Hypotheses: $H_0: \mu_1 = \mu_2 = \ldots = \mu_9$
    $H_a:$ Not all means are equal

Decision Rule: Reject H_0 if $F > 2.02$ $F(0.05; 8, 103) \approx 2.02$
    Do not reject H_0 if $F \leq 2.02$

Test Statistic: $F = 14.82$

OR

Decision Rule: Reject H_0 if p value < 0.05
    Do not reject H_0 if p value ≥ 0.05

Test Statistic: p value $= 0.000$

Decision: Reject H_0

Conclusion: There is a difference in the average number of collisions for different types of car.

9.3   First Test for Interaction Effects

Hypotheses: $H_0:$ No interaction between factors A and B exists
    $H_a:$ Factors A and B interact

Decision Rule: Reject H_0 if $F > 3.00$ $F(0.05; 6, 12) = 3.00$
    Do not reject H_0 if $F \leq 3.00$

Test Statistic: $F = 0.20$

OR

Decision Rule: Reject H_0 if p value < 0.05
    Do not reject H_0 if p value ≥ 0.05

Test Statistic: p value $= 0.970$

Decision: Do not reject H_0

Conclusion: There are no interaction effects. Note that this means that the F tests for main effects can be used.

Test for Main Effects Due to Industry

Hypotheses: H_0: $\alpha_1 = \alpha_2 = \alpha_3 = \alpha_4$
H_a: The α_i are not all equal

Decision Rule: Reject H_0 if $F > 3.49$ $F(0.05; 3, 12) = 3.49$
Do not reject H_0 if $F \leq 3.49$

Test Statistic: $F = 49.82$

OR

Decision Rule: Reject H_0 if p value < 0.05
Do not reject H_0 if p value ≥ 0.05

Test Statistic: p value $= 0.000$

Decision: Reject H_0

Conclusion: The α_i are not all equal. There are differences in treatment means associated with different industries.

Test for Main Effects Due to Contact

Hypotheses: H_0: $\beta_1 = \beta_2 = \beta_3$
H_a: The β_i are not all equal

Decision Rule: Reject H_0 if $F > 3.89$ $F(0.05; 2, 12) = 3.89$
Do not reject H_0 if $F \leq 3.89$

Test Statistic: $F = 0.03$

OR

Decision Rule: Reject H_0 if p value < 0.05
Do not reject H_0 if p value ≥ 0.05

Test Statistic: p value $= 0.974$

Decision: Do not reject H_0

Conclusion: The β_i are equal. There are no differences in treatment means associated with different contacts.

9.5 Hypotheses: $H_0: \mu_1 = \mu_2 = \mu_3$
H_a: Not all means are equal

Decision Rule: Reject H_0 if $F > 3.68$ $F(0.05; 2, 15) = 3.68$
Do not reject H_0 if $F \leq 3.68$

Test Statistic: $F = 102.32$

OR

Decision Rule: Reject H_0 if p value < 0.05
Do not reject H_0 if p value ≥ 0.05

Test Statistic: p value $= 0.000$

Decision: Reject H_0

Conclusion: There is a difference in the average production rates.

ANOVA Output

```
Source   DF        SS       MS       F       P
LINE      2    1357.44   678.72   102.32   0.000
Error    15      99.50     6.63
Total    17    1456.94

S = 2.576    R-Sq = 93.17%    R-Sq(adj) = 92.26%


                          Individual 95% CIs For Mean Based on
                          Pooled StDev
Level   N    Mean    StDev   ----+---------+---------+---------+----
1       6   36.167   1.472   (--*---)
2       6   54.333   3.141                                    (---*--)
3       6   35.667   2.805   (--*---)
                             ----+---------+---------+---------+----
                              36.0      42.0      48.0      54.0

Pooled StDev = 2.576
```

9.7 **a** Hypotheses: $H_0: \gamma_1 = \gamma_2 = \gamma_3$
H_a: Not all γ_k are equal

Decision Rule: Reject H_0 if $F > 6.94$ $\quad\quad F(0.05; 2, 4) = 6.94$
Do not reject H_0 if $F \leq 6.94$

Test Statistic: $F = 103.32$

OR

Decision Rule: Reject H_0 if p value < 0.05
Do not reject H_0 if p value ≥ 0.05

Test Statistic: p value $= 0.000$

Decision: Reject H_0

Conclusion: There is a difference in the average test scores due to training program.

b $\mu_1 - \mu_2$: $(82.33 - 96.00) \pm 2.776(1.45)\sqrt{\dfrac{2}{3}}$ or $(-16.96, -10.38)$

<u>ANOVA Output</u>

```
Source    DF       SS        MS        F       P
PROGRAM    2   436.222   218.111   103.32   0.000
BLOCK      2     2.889     1.444     0.68   0.555
Error      4     8.444     2.111
Total      8   447.556

S = 1.453    R-Sq = 98.11%    R-Sq(adj) = 96.23%
```

9.9 To determine whether average sales differ on different days of the week, the following hypotheses should be tested:

Hypotheses: $H_0: \mu_1 = \mu_2 = \ldots = \mu_7$
H_a: Not all means are equal

Decision Rule: Reject H_0 if $F > 2.34$ $F(0.05; 6, 49) \approx 2.34$
Do not reject H_0 if $F \leq 2.34$

Test Statistic: $F = 29.21$

OR

Decision Rule: Reject H_0 if p value < 0.05
Do not reject H_0 if p value ≥ 0.05

Test Statistic: p value = 0.000

Decision: Reject H_0

Conclusion: There is a difference in the average sales on different days of the week

Although the individual 95% confidence intervals printed out by MINITAB are not the best way to compare the means, it is pretty obvious in this case that average sales on Friday and Saturday are much higher than on the other days of the week. Using all pairwise comparisons, it can be shown that the average sales for Monday, Tuesday, Wednesday, Thursday, and Sunday are not significantly different from each other. However, the average sales for Friday and Saturday do differ significantly from any other day (although not from each other). Here is an example of one of the pairwise comparison intervals using the Bonferroni method. This one compares Thursday and Friday:

$\mu_{Thurs} - \mu_{Fri}$: $(3217.9 - 6465.2) \pm 3.20(981.2)\sqrt{\dfrac{1}{8} + \dfrac{1}{8}}$

or   (-4817.22, -16776.38)

Note that the negative signs indicate that Friday sales are greater than Thursday sales. The t value of 3.20 used in the interval is the Bonferroni t value for a 95% familywise confidence interval, assuming we want to make all pairwise comparisons. There will be 21 possible familywise comparisons of the seven means. The t value used is the t value with $\alpha = 0.05/42 = 0.00119$ and 49 degrees of freedom. The t value was determined using MINITAB.

ANOVA Output

```
Source  DF         SS         MS       F       P
DAY      6  168697125   28116187   29.21   0.000
Error   49   47171302     962680
Total   55  215868427
```

S = 981.2   R-Sq = 78.15%   R-Sq(adj) = 75.47%

```
                                 Individual 95% CIs For Mean Based on Pooled
                                 StDev
Level  N    Mean    StDev    -+---------+---------+---------+--------
1      8   2107.2   263.0     (----*----)
2      8   2745.6   636.3          (---*----)
3      8   2889.3   565.9           (---*----)
4      8   3217.9  1008.2             (---*----)
5      8   6465.2  1476.4                               (----*----)
6      8   6621.5  1581.3                                (----*----)
7      8   2856.4   497.7           (----*----)
                              -+---------+---------+---------+--------
                             1500      3000      4500      6000
```

Pooled StDev = 981.2

9.11 **a** Hypotheses: $H_0: \mu_1 = \mu_2 = \mu_3 = \mu_4$
H_a: Not all means are equal

Decision Rule: Reject H_0 if $F > 4.07$ F(0.05; 3, 8) = 4.07
Do not reject H_0 if $F \leq 4.07$

Test Statistic: $F = 14.00$

OR

Decision Rule: Reject H_0 if p value < 0.05
Do not reject H_0 if p value ≥ 0.05

Test Statistic: p value = 0.002

Decision: Reject H_0

Conclusion: There is a difference in the average lifetimes.

b $\mu_1 - \mu_2$: $(99.00 - 100.00) \pm 2.306(1.00)\sqrt{\frac{1}{3} + \frac{1}{3}}$ or   (-2.88, 0.88)

c   See MINITAB output

ANOVA Output

```
Source     DF     SS      MS      F       P
SUPPLIER    3   42.00   14.00   14.00   0.002
Error       8    8.00    1.00
Total      11   50.00

S = 1    R-Sq = 84.00%    R-Sq(adj) = 78.00%
```

```
                           Individual 90% CIs For Mean Based on
                           Pooled StDev
Level   N    Mean   StDev   -----+---------+---------+---------+----
1       3    99.00   1.00       (----*----)
2       3   100.00   1.00            (----*----)
3       3   103.00   1.00                          (----*----)
4       3    98.00   1.00   (----*----)
                            -----+---------+---------+---------+----
                                98.0      100.0     102.0     104.0
```

Pooled StDev = 1.00

Tukey 95% Simultaneous Confidence Intervals
All Pairwise Comparisons among Levels of SUPPLIER

Individual confidence level = 98.74%

SUPPLIER = 1 subtracted from:

```
SUPPLIER   Lower    Center    Upper   ---------+---------+---------+---------+
2         -1.615    1.000    3.615              (------*-----)
3          1.385    4.000    6.615                       (------*------)
4         -3.615   -1.000    1.615        (------*-----)
                                        ---------+---------+---------+---------+
                                              -4.0       0.0       4.0       8.0
```

SUPPLIER = 2 subtracted from:

```
SUPPLIER   Lower    Center    Upper   ---------+---------+---------+---------+
3          0.385    3.000    5.615                       (------*-----)
4         -4.615   -2.000    0.615          (------*------)
                                        ---------+---------+---------+---------+
                                              -4.0       0.0       4.0       8.0
```

SUPPLIER = 3 subtracted from:

```
SUPPLIER   Lower    Center    Upper   ---------+---------+---------+---------+
4         -7.615   -5.000   -2.385     (------*-----)
                                        ---------+---------+---------+---------+
                                              -4.0       0.0       4.0       8.0
```

CHAPTER 10
Qualitative Dependent Variables: An Introduction to Discriminant Analysis and Logistic Regression

10.1   **a** $H_0: \beta_1 = 0.0$
$H_a: \beta_1 \neq 0.0$

Decision Rule: Reject H_0 if $z > 1.96$ or $z < -1.96$
Do not reject H_0 if $-1.96 \leq z \leq 1.96$

Test Statistic: $z = 3.06$

OR

Decision Rule: Reject H_0 if p value < 0.05
Do not reject H_0 if p value ≥ 0.05

Test Statistic: p value $= 0.002$

Decision: Reject H_0

Conclusion: TEST1 is useful

$H_0: \beta_2 = 0.0$
$H_a: \beta_2 \neq 0.0$

Decision Rule: Reject H_0 if $z > 1.96$ or $z < -1.96$
Do not reject H_0 if $-1.96 \leq z \leq 1.96$

Test Statistic: $z = 1.62$

OR

Decision Rule: Reject H_0 if p value < 0.05
Do not reject H_0 if p value ≥ 0.05

Test Statistic: p value $= 0.106$

Decision: Do not reject H_0

Conclusion: TEST2 is not useful

b Use the output in Figure 10.6. The results of TEST2 do not appear useful in helping to predict success for these employees.

1. -43.37 + 0.4897(94) = 2.6618   probability 0.93
2. -43.37 + 0.4897(80) = -4.194   probability 0.01
3. -43.37 + 0.4897(82) = -3.2146   probability 0.04
4. -43.37 + 0.4897(90) = 0.703   probability 0.67

Potential employees 1 and 4 would be classified in the "success" group, while numbers 2 and 3 would not.

10.3

Logistic Regression Output

```
Link Function: Logit

Response Information

Variable   Value   Count
PURCHASE   1         18    (Event)
           0         22
           Total     40


Logistic Regression Table

                                                    Odds      95% CI
Predictor     Coef      SE Coef      Z       P      Ratio   Lower   Upper
Constant    -8.15481    2.74176    -2.97   0.003
INCOME       0.109872   0.0387490   2.84   0.005    1.12    1.03    1.20
AGE          1.21173    0.513574    2.36   0.018    3.36    1.23    9.19

Log-Likelihood = -19.373
Test that all slopes are zero: G = 16.304, DF = 2, P-Value = 0.000

Goodness-of-Fit Tests

Method            Chi-Square   DF      P
Pearson            29.1859     31    0.560
Deviance           33.2014     31    0.360
Hosmer-Lemeshow     6.2192      8    0.623

Table of Observed and Expected Frequencies:
(See Hosmer-Lemeshow Test for the Pearson Chi-Square Statistic)

                              Group
Value    1     2     3     4     5     6     7     8     9    10    Total
1
  Obs    0     0     1     2     1     3     1     3     3     4     18
  Exp   0.3   0.6   0.9   1.1   1.2   1.6   2.1   2.8   3.5   3.9
0
  Obs    4     4     3     2     3     1     3     1     1     0     22
  Exp   3.7   3.4   3.1   2.9   2.8   2.4   1.9   1.2   0.5   0.1
Total    4     4     4     4     4     4     4     4     4     4     40


Measures of Association:
(Between the Response Variable and Predicted Probabilities)

Pairs         Number   Percent   Summary Measures
Concordant      328     82.8     Somers' D                0.68
Discordant       60     15.2     Goodman-Kruskal Gamma    0.69
Ties              8      2.0     Kendall's Tau-a          0.34
Total           396    100.0
```

$H_0: \beta_1 = 0.0$
$H_a: \beta_1 \neq 0.0$

Decision Rule: Reject H_0 if $z > 1.96$ or $z < -1.96$
Do not reject H_0 if $-1.96 \leq z \leq 1.96$

Test Statistic: $z = 2.84$

OR

Decision Rule: Reject H_0 if p value < 0.05
Do not reject H_0 if p value ≥ 0.05

Test Statistic: p value $= 0.005$

Decision: Reject H_0

Conclusion: INCOME is useful

$H_0: \beta_2 = 0.0$
$H_a: \beta_2 \neq 0.0$

Decision Rule: Reject H_0 if $z > 1.96$ or $z < -1.96$
Do not reject H_0 if $-1.96 \leq z \leq 1.96$

Test Statistic: $z = 2.36$

OR

Decision Rule: Reject H_0 if p value < 0.05
Do not reject H_0 if p value ≥ 0.05

Test Statistic: p value $= 0.018$

Decision: Reject H_0

Conclusion: AGE is useful

10.5

Logistic Regression Output

```
Link Function: Logit

Response Information

Variable  Value  Count
WIN       1      51    (Event)
          0      12
          Total  63
```

Logistic Regression Table

```
                                                      95% CI
Predictor       Coef     SE Coef      Z       P  Odds Ratio    Lower        Upper
Constant    -3.15305     4.67149  -0.67   0.500
DIFF        -0.452871    0.168326 -2.69   0.007        0.64     0.46         0.88
PCTHIGH      6.46083     5.11392   1.26   0.206      639.59     0.03  14419852.43
PCTLOW      -2.86870     4.21713  -0.68   0.496        0.06     0.00       220.71

Log-Likelihood = -22.700
Test that all slopes are zero: G = 15.951, DF = 3, P-Value = 0.001
```

Goodness-of-Fit Tests

```
Method           Chi-Square   DF      P
Pearson             80.6826   58  0.026
Deviance            45.4003   58  0.886
Hosmer-Lemeshow     24.5018    8  0.002
```

Table of Observed and Expected Frequencies:
(See Hosmer-Lemeshow Test for the Pearson Chi-Square Statistic)

```
                              Group
Value   1     2     3     4     5     6     7     8     9    10   Total
1
  Obs   3     1     5     6     5     6     7     6     5     7     51
  Exp  2.6   3.2   3.8   5.3   5.2   5.5   6.6   5.9   6.0   7.0
0
  Obs   3     5     1     1     1     0     0     0     1     0     12
  Exp  3.4   2.8   2.2   1.7   0.8   0.5   0.4   0.1   0.0   0.0
Total   6     6     6     7     6     6     7     6     6     7     63
```

Measures of Association:
(Between the Response Variable and Predicted Probabilities)

```
Pairs        Number  Percent  Summary Measures
Concordant      516    84.3   Somers' D               0.69
Discordant       95    15.5   Goodman-Kruskal Gamma   0.69
Ties              1     0.2   Kendall's Tau-a         0.22
Total           612   100.0
```

$H_0: \beta_1 = 0.0$
$H_a: \beta_1 \neq 0.0$

Decision Rule: Reject H_0 if $z > 1.96$ or $z < -1.96$
Do not reject H_0 if $-1.96 \leq z \leq 1.96$

Test Statistic: $z = -2.69$

OR

Decision Rule: Reject H_0 if p value < 0.05
Do not reject H_0 if p value ≥ 0.05

Test Statistic: p value $= 0.007$

Decision: Reject H_0

Conclusion: DIFF is useful

$H_0: \beta_2 = 0.0$
$H_a: \beta_2 \neq 0.0$

Decision Rule: Reject H_0 if $z > 1.96$ or $z < -1.96$
Do not reject H_0 if $-1.96 \leq z \leq 1.96$

Test Statistic: $z = 1.26$

OR

Decision Rule: Reject H_0 if p value < 0.05
Do not reject H_0 if p value ≥ 0.05

Test Statistic: p value $= 0.206$

Decision: Do not reject H_0

Conclusion: PCTHIGH is not useful

H_0: $\beta_3 = 0.0$
H_a: $\beta_3 \neq 0.0$

Decision Rule: Reject H_0 if $z > 1.96$ or $z < -1.96$
Do not reject H_0 if $-1.96 \leq z \leq 1.96$

Test Statistic: $z = -0.68$

OR

Decision Rule: Reject H_0 if p value < 0.05
Do not reject H_0 if p value ≥ 0.05

Test Statistic: p value $= 0.496$

Decision: Do not reject H_0

Conclusion: PCTLOW is not useful

CHAPTER 11
Forecasting Methods for Time Series Data

11.1

METHOD	MAPE	MAD	MSD
Moving Average Length 1	26	47021	3301636525
Moving Average Length 2	25	44821	2825426091
Moving Average Length 3	23	40870	2537149730
Moving Average Length 4	21	36520	2203979413
Moving Average Length 5	20	35561	2024066925
Moving Average Length 6	20	35983	2112865326
Smoothing Constant Alpha 0.0083969	20	34082	1822769972

In-sample accuracy measures are used to compare methods. No out-of-sample forecasts were generated. The best choice for WestCo to forecast daily envelope sales at this point appears to be simply to use the average of past sales. The naïve method (moving average length 1) is worst among the methods examined. The optimal smoothing value for exponential smoothing found by MINITAB is close to zero suggesting that nearly equal weight be put on all past observations. This suggests that the observations have very little systematic pattern (are close to random) and that the average would be a good candidate to generate future forecasts. (The average does have the property that it is the single value that will minimize the sum of squared forecast errors.)

11.3

Period	SES Forecast	DES Forecast	Actual
241	5.99787	6.01920	5.8
242	5.99787	6.06493	5.9
243	5.99787	6.11065	5.8
244	5.99787	6.15638	6.0
245	5.99787	6.20210	6.1
246	5.99787	6.24783	6.3
247	5.99787	6.29355	6.2
248	5.99787	6.33927	6.1
249	5.99787	6.38500	6.1
250	5.99787	6.43072	6.0
251	5.99787	6.47645	5.9
252	5.99787	6.52217	5.7

Using the standard error measures applied to the out-of-sample data, SES forecasts are more accurate than DES. Note that the conclusions from the out-of-sample results differ from those based on in-sample accuracy statistics.

	MAD	MSD	MAPE
SES	0.142	0.029	2.38%
DES	0.288	0.129	4.88%

11.5 In-sample accuracy measures are used to compare methods. No out-of-sample forecasts were generated. Single and double exponential smoothing were considered. Winter's method was also considered because it was unclear if there was seasonal variation in the data. The multiplicative version of Winter's method was used with the default smoothing constants (0.2, 0.2, 0.2). Using the standard error measures, SES forecasts are slightly more accurate than DES and Winter's Method.

	MAD	MSD	MAPE
SES	459	389494	17%
DES	473	431732	17%
Winter's	487	420359	18%

11.7 SES, DES and the naïve forecast (MA with length 1) were considered.

Period	SES Forecast	DES Forecast	Naïve Forecast	Actual
181	4.27468	4.15576	4.25	4.25
182	4.27468	4.04731	4.25	4.25
183	4.27468	3.93886	4.25	4.25
184	4.27468	3.83041	4.25	4.25
185	4.27468	3.72197	4.25	4.25
186	4.27468	3.61352	4.25	4.22
187	4.27468	3.50507	4.25	4.00
188	4.27468	3.39662	4.25	4.00
189	4.27468	3.28817	4.25	4.00
190	4.27468	3.17972	4.25	4.00
191	4.27468	3.07128	4.25	4.00
192	4.27468	2.96283	4.25	4.00

Using the standard error measures applied to out-of-sample forecasts, the naive forecasts are slightly more accurate than SES. Both SES and naïve are considerably more accurate than DES. Note that the conclusions from the out-of-sample results differ from those based on in-sample accuracy statistics.

	MAD	MSD	MAPE
SES	0.152	0.038	3.78%
DES	0.563	0.391	13.82%
NAÏVE	0.128	0.031	3.18%

11.9 Because of the strong seasonal variation, Winter's method (additive and multiplicative) and decomposition (additive and multiplicative) were considered. Using the standard error measures, Additive Decomposition was slightly more accurate for the in-sample time period. No out-of-sample forecasts were generated. Only the default smoothing parameter values (0.2, 0.2, 0.2) were examined for the Winter's Methods.

	MAD	MSD	MAPE
Winter's M	2.7456	12.6723	4.5996%
Winter's A	2.6333	12.3410	4.4866%
Decomp M	2.26314	9.29202	3.91370%
Decomp A	2.22837	9.01055	3.86912%

Here are forecasts for the next two years from each of the four methods considered.

Period	Winter's M	Winter's A	Decomp M	Decomp A
301	45.8133	45.4822	45.9369	46.1735
302	50.3134	50.0853	51.0550	51.0843
303	55.2559	55.1475	57.6510	57.6451
304	64.9837	64.9594	66.3716	66.2142
305	72.8591	73.0611	74.0229	73.7438
306	79.7018	80.0838	81.8183	81.3838
307	84.3220	84.8471	87.0331	86.6737
308	84.2360	84.8237	86.5812	86.2137
309	76.7098	77.1581	79.4158	78.9161
310	65.6835	65.8807	68.1180	67.9144
311	54.4978	54.4913	56.4268	56.3982
312	46.1306	45.9848	49.0209	48.9944
313	45.2989	45.0965	46.0062	46.2531
314	49.7480	49.6996	51.1320	51.1639
315	54.6343	54.7618	57.7379	57.7247
316	64.2520	64.5737	66.4717	66.2938
317	72.0380	72.6755	74.1345	73.8233
318	78.8027	79.6982	81.9416	81.4633
319	83.3699	84.4614	87.1644	86.7533
320	83.2840	84.4381	86.7117	86.2932
321	75.8420	76.7725	79.5355	78.9957
322	64.9398	65.4951	68.2206	67.9940
323	53.8802	54.1057	56.5118	56.4777
324	45.6072	45.5992	49.0948	49.0739

APPENDIX A
Summation Notation

a   66

b   854

c   0

d   128

e   132

APPENDIX D
Matrices and Their Application to Regression Analysis

1      $\mathbf{A} + \mathbf{B} = \begin{bmatrix} 5 & 4 & 10 \\ 4 & 2 & 7 \\ 4 & 5 & 8 \end{bmatrix}$

2      $\mathbf{A} + \mathbf{B} = \begin{bmatrix} 5 & 7 \\ 8 & 15 \end{bmatrix}$

3      $\mathbf{A} + \mathbf{B} = \begin{bmatrix} 7 \\ 4 \\ 7 \end{bmatrix}$

4      $\mathbf{A}^T = \begin{bmatrix} 1 & 2 & 3 \\ 3 & 1 & 1 \\ 4 & 2 & 5 \end{bmatrix}$

5     $\mathbf{B}^T = \begin{bmatrix} 1 & 3 \\ 7 & 4 \\ 5 & 6 \end{bmatrix}$

6     $\mathbf{C}^T = \begin{bmatrix} 1 & 3 & 2 \end{bmatrix}$

7     $\mathbf{AB} = \begin{bmatrix} 6 & 23 \\ 34 & 89 \end{bmatrix}$

8     $\mathbf{AB} = \begin{bmatrix} 11 & 20 & 33 \\ 12 & 11 & 23 \\ 19 & 24 & 38 \end{bmatrix}$

9     $\mathbf{AA}^T = \begin{bmatrix} 75 & 61 \\ 61 & 61 \end{bmatrix}$

10   **AB = A** because **B** is the identity matrix

11     $\mathbf{A}^{-1} = \begin{bmatrix} -3 & 1 \\ 2 & -0.5 \end{bmatrix}$

12     $\mathbf{A}^{-1} = \begin{bmatrix} 0.3 & -0.1 \\ -0.2 & 0.4 \end{bmatrix}$

13     $\mathbf{X} = \begin{bmatrix} 1 & 1 \\ 1 & 2 \\ 1 & 4 \\ 1 & 5 \\ 1 & 6 \end{bmatrix} \quad \mathbf{Y} = \begin{bmatrix} 5 \\ 6 \\ 9 \\ 10 \\ 14 \end{bmatrix}$

14 $\mathbf{X}^T = \begin{bmatrix} 1 & 1 & 1 & 1 & 1 \\ 1 & 2 & 4 & 5 & 6 \end{bmatrix}$

$\mathbf{X}^T\mathbf{X} = \begin{bmatrix} 5 & 18 \\ 18 & 82 \end{bmatrix}$

$(\mathbf{X}^T\mathbf{X})^{-1} = \begin{bmatrix} 84/86 & -18/86 \\ -18/86 & 5/86 \end{bmatrix}$

$\mathbf{X}^T\mathbf{Y} = \begin{bmatrix} 44 \\ 187 \end{bmatrix}$

$\mathbf{b} = (\mathbf{X}^T\mathbf{X})^{-1}(\mathbf{X}^T\mathbf{Y}) = \begin{bmatrix} 330/86 \\ 143/86 \end{bmatrix}$ OR $\mathbf{b} = \begin{bmatrix} 3.837 \\ 1.663 \end{bmatrix}$

15 $SSE = \mathbf{Y}^T\mathbf{Y} - \mathbf{b}^T(\mathbf{X}^T\mathbf{Y}) = 3.24$

$s_e = \sqrt{\dfrac{SSE}{n-2}} = \sqrt{\dfrac{3.24}{3}} = 1.04$